Handbook of Exact String Matching Algorithms

Texts in Algorithms

Series Editors: A. M. Gibbons and C. S. Iliopoulos
King's College London

Handbook of Exact String Matching Algorithms

Christian Charras

and

Thierry Lecroq

© Individual author and King's College 2004. All rights reserved.

ISBN 0-9543006-4-5
King's College Publications
Scientific Director: Dov Gabbay
Managing Director: Jane Spurr
Department of Computer Science
Strand, London WC2R 2LS, UK
kcp@dcs.kcl.ac.uk

Cover design by Richard Fraser, www.avalonarts.co.uk
Printed by Lightning Source, Milton Keynes, UK

All rights reserved. No part of this publication may be reproduced, stored in a retrieval system or transmitted, in any form, or by any means, electronic, mechanical, photocopying, recording or otherwise, without prior permission, in writing, from the publisher.

CONTENTS

LIST OF FIGURES   xiii

PREFACE   xv

CHAPTER 1   INTRODUCTION   1
1   From left to right   2
2   From right to left   3
3   In a specific order   3
4   In any order   4
5   Conventions   4
5.1   Definitions   4
5.2   Implementations   5

CHAPTER 2   BRUTE FORCE ALGORITHM   9
1   Main features   9
2   Description   9
3   The C code   10
4   The example   10

CHAPTER 3   SEARCH WITH AN AUTOMATON   15
1   Main features   15
2   Description   15
3   The C code   16
4   The example   16
5   References   20

CHAPTER 4   KARP-RABIN ALGORITHM   21
1   Main features   21
2   Description   21
3   The C code   22
4   The example   23
5   References   26

CHAPTER 5 SHIFT OR ALGORITHM — 27
1 Main features — 27
2 Description — 27
3 The C code — 28
4 The example — 29
5 References — 30

CHAPTER 6 MORRIS-PRATT ALGORITHM — 31
1 Main Features — 31
2 Description — 31
3 The C code — 32
4 The example — 33
5 References — 35

CHAPTER 7 KNUTH-MORRIS-PRATT ALGORITHM — 37
1 Main Features — 37
2 Description — 37
3 The C code — 38
4 The example — 39
5 References — 41

CHAPTER 8 SIMON ALGORITHM — 45
1 Main features — 45
2 Description — 45
3 The C code — 46
4 The example — 48
5 References — 51

CHAPTER 9 COLUSSI ALGORITHM — 53
1 Main features — 53
2 Description — 53
3 The C code — 55
4 The example — 58
5 References — 60

CHAPTER 10 GALIL-GIANCARLO ALGORITHM — 61
1 Main features — 61
2 Description — 61
3 The C code — 62
4 The example — 63
5 References — 65

CHAPTER 11 APOSTOLICO-CROCHEMORE ALGORITHM 67
1. Main features 67
2. Description 67
3. The C code 68
4. The example 69
5. References 71

CHAPTER 12 NOT SO NAIVE ALGORITHM 73
1. Main features 73
2. Description 73
3. The C code 73
4. The example 74
5. References 77

CHAPTER 13 FORWARD DAWG MATCHING ALGORITHM 79
1. Main Features 79
2. Description 79
3. The C code 80
4. The example 81
5. References 82

CHAPTER 14 BOYER-MOORE ALGORITHM 83
1. Main Features 83
2. Description 83
3. The C code 85
4. The example 87
5. References 88

CHAPTER 15 GALIL ALGORITHM 91
1. Main Features 91
2. Description 91
3. The C code 91
4. The example 92
5. References 93

CHAPTER 16 SMYTH ALGORITHM 95
1. Main Features 95
2. Description 95
3. The C code 95
4. The example 96
5. References 97

CHAPTER 17 TURBO-BM ALGORITHM 99
1 Main Features 99
2 Description 99
3 The C code 100
4 The example 102
5 References 103

CHAPTER 18 APOSTOLICO-GIANCARLO ALGORITHM 105
1 Main Features 105
2 Description 105
3 The C code 108
4 The example 109
5 References 110

CHAPTER 19 REVERSE COLUSSI ALGORITHM 113
1 Main features 113
2 Description 113
3 The C code 114
4 The example 116
5 References 118

CHAPTER 20 HORSPOOL ALGORITHM 119
1 Main Features 119
2 Description 119
3 The C code 119
4 The example 120
5 References 121

CHAPTER 21 FAST SEARCH ALGORITHM 123
1 Main Features 123
2 Description 123
3 The C code 124
4 The example 124
5 References 126

CHAPTER 22 QUICK SEARCH ALGORITHM 127
1 Main Features 127
2 Description 127
3 The C code 128
4 The example 128
5 References 129

CHAPTER 23 TURBO SEARCH ALGORITHM   131
1 Main Features 131
2 Description 131
3 The C code 131
4 The example 132
5 References 134

CHAPTER 24 TUNED BOYER-MOORE ALGORITHM   135
1 Main Features 135
2 Description 135
3 The C code 136
4 The example 136
5 References 138

CHAPTER 25 ZHU-TAKAOKA ALGORITHM   139
1 Main features 139
2 Description 139
3 The C code 140
4 The example 141
5 References 142

CHAPTER 26 BERRY-RAVINDRAN ALGORITHM   143
1 Main features 143
2 Description 143
3 The C code 144
4 The example 144
5 References 146

CHAPTER 27 SMITH ALGORITHM   147
1 Main features 147
2 Description 147
3 The C code 147
4 The example 148
5 References 149

CHAPTER 28 RAITA ALGORITHM   151
1 Main features 151
2 Description 151
3 The C code 152
4 The example 152
5 References 154

CHAPTER 29 REVERSE FACTOR ALGORITHM 155
1. Main Features 155
2. Description 155
3. The C code 156
4. The example 159
5. References 160

CHAPTER 30 TURBO REVERSE FACTOR ALGORITHM 161
1. Main Features 161
2. Description 161
3. The C code 162
4. The example 164
5. References 165

CHAPTER 31 BACKWARD SUFFIX ORACLE MATCHING ALGORITHM 167
1. Main Features 167
2. Description 167
3. The C code 168
4. The example 171
5. References 172

CHAPTER 32 BACKWARD NONDETERMINISTIC DAWG MATCHING ALGORITHM 173
1. Main Features 173
2. Description 173
3. The C code 174
4. The example 174
5. References 177

CHAPTER 33 GALIL-SEIFERAS ALGORITHM 179
1. Main features 179
2. Description 179
3. The C code 180
4. The example 183
5. References 185

CHAPTER 34 TWO WAY ALGORITHM 187
1. Main features 187
2. Description 187
3. The C code 188
4. The example 191

5	References	193

CHAPTER 35 STRING MATCHING ON ORDERED ALPHABETS 195
1	Main features	195
2	Description	195
3	The C code	196
4	The example	199
5	References	200

CHAPTER 36 OPTIMAL MISMATCH ALGORITHM 201
1	Main features	201
2	Description	201
3	The C code	201
4	The example	204
5	References	205

CHAPTER 37 MAXIMAL SHIFT ALGORITHM 207
1	Main features	207
2	Description	207
3	The C code	207
4	The example	209
5	References	210

CHAPTER 38 SKIP SEARCH ALGORITHM 211
1	Main features	211
2	Description	211
3	The C code	211
4	The example	213
5	References	214

CHAPTER 39 KMPSKIP SEARCH ALGORITHM 215
1	Main features	215
2	Description	215
3	The C code	216
4	The example	219
5	References	220

CHAPTER 40 ALPHA SKIP SEARCH ALGORITHM 221
1	Main features	221
2	Description	221
3	The C code	221
4	The example	224

5   References 225

APPENDIX I   EXAMPLE OF GRAPH IMPLEMENTATION   227

INDEX   235

LIST OF FIGURES

5.1 Meaning of vector R_j in the Shift-Or algorithm. 28
6.1 Shift in the Morris-Pratt algorithm: v is the border of u. 32
7.1 Shift in the Knuth-Morris-Pratt algorithm: v is a border of u and $a \neq c$. 38
9.1 Mismatch with a nohole. Noholes are black circles and are compared from left to right. In this situation, after the shift, it is not necessary to compare the first two noholes again. 54
9.2 Mismatch with a hole. Noholes are black circles and are compared from left to right while holes are white circles and are compared from right to left. In this situation, after the shift, it is not necessary to compare the matched prefix of the pattern again. 55
11.1 At each attempt of the Apostolico-Crochemore algorithm we consider a triple (i, j, k). 67
14.1 The good-suffix shift, u re-occurs preceded by a character c different from a. 84
14.2 The good-suffix shift, only a suffix of u re-occurs in x. 84
14.3 The bad-character shift, b occurs in x. 84
14.4 The bad-character shift, b does not occur in x. 85
17.1 A turbo-shift can apply when $|v| < |u|$. 100
17.2 $c \neq d$ so they cannot be aligned with the same character in v. . . 100
18.1 Case 1, $k > \text{suff}[i]$ and $\text{suff}[i] = i + 1$, an occurrence of x is found. 106
18.2 Case 2, $k > \text{suff}[i]$ and $\text{suff}[i] \leq i$, a mismatch occurs between $y[i + j - \text{suff}[i]]$ and $x[i - \text{suff}[i]]$. 106
18.3 Case 3, $k < \text{suff}[i]$ a mismatch occurs between $y[i + j - k]$ and $x[i - k]$. 106
18.4 Case 4, $k = \text{suff}[i]$ and $a \neq b$. 107
30.1 Impossible overlap if z is an acyclic word. 162
33.1 A perfect factorization of x. 180
35.1 Typical attempt during the String Matching on Ordered Alphabets algorithm. 195
35.2 Function nextMaximalSuffix: meaning of the variables i, j, k and p. 196

39.1 General situation during the searching phase of the KmpSkip algorithm. 216

PREFACE

The material presented in this book first started as a small collection of Xwindows applications. They were designed in order to teach string matching to Masters students in computer science at the University of Rouen. The advent of the World Wide Web and the enormous possibilities offered by Java applets enabled us to launch a web site dedicated to exact string matching in 1995. It was first hosted by the web site of the Computer Science Department of the University of Rouen. Since 1997 it has been hosted by the web site of the Gaspard Monge Institute of the University of Marne-la-Vallée at the following URL:
http://www-igm.univ-mlv.fr/~lecroq/string.

We are grateful for Maxime Crochemore to have made this possible.

We would also like to thank Costas Iliopoulos for making it possible to publish this collection of algorithms in a book.

Many thanks, as well, to Christophe Hancart who designed the LaTeX style files that we used for the examples and to Jane Spurr for her help.

And finally, thank you to all the users of our web site who notify us of errors and provide comments.

CHRISTIAN CHARRAS
THIERRY LECROQ
Mont-Saint-Aignan
September, 2003

CHAPTER 1

INTRODUCTION

String matching is a very important subject in the wider domain of text processing. String matching algorithms are basic components used in implementations of practical softwares existing under most operating systems. Moreover, they emphasize programming methods that serve as paradigms in other fields of computer science (system or software design). Finally, they also play an important role in theoretical computer science by providing challenging problems.

Although data are memorized in various ways, text remains the main form to exchange information. This is particularly evident in literature or linguistics where data are composed of huge corpora and dictionaries. This applies as well to computer science where a large amount of data are stored in linear files. And this is also the case, for instance, in molecular biology since biological molecules can often be approximated as sequences of nucleotides or amino acids. Furthermore, the quantity of available data in these fields tend to double every eighteen months. This is the reason why algorithms should be efficient even if the speed and capacity of storage of computers increase regularly.

String matching consists in finding one, or more generally, all the occurrences of a string (more generally called a **pattern**) in a **text**. All the algorithms in this book output all occurrences of the pattern in the text. The pattern is denoted by $x = x[0 \mathinner{.\,.} m-1]$; its length is equal to m. The text is denoted by $y = y[0 \mathinner{.\,.} n-1]$; its length is equal to n. Both strings are build over a finite set of character called an **alphabet** denoted by Σ with size is equal to σ.

Applications require two kinds of solution depending on which string, the pattern or the text, is given first. Algorithms based on the use of automata or combinatorial properties of strings are commonly implemented to preprocess the pattern and solve the first kind of problem. The notion of indexes realized by trees or automata is used in the second kind of solutions. This book will only investigate algorithms of the first kind.

String matching algorithms of the present book work as follows. They scan the text with the help of a **window** which size is generally equal to m. They first align the left ends of the window and the text, then compare the characters of the window with the characters of the pattern — this specific work is called an **attempt** — and after a whole match of the pattern or after a mismatch they **shift** the window to the right. They repeat the same procedure again until the right end of the window goes beyond the right end of the text. This mechanism is usually called the **sliding window mechanism**. We associate each attempt with the positions j and $j+m-1$ in the text when the window is positioned on $y[j \mathinner{.\,.} j+m-1]$: we say that the attempt is at the left position j and at the right position $j+m-1$.

The brute force algorithm (see chapter 2) locates all occurrences of x in y in time $O(m \times n)$. The many improvements of the brute force method can be classified depending on the order they performed the comparisons between pattern characters and text characters et each attempt. Four categories arise: the most natural way to perform the comparisons is from left to right, which is the reading direction; performing the comparisons from right to left generally leads to the best algorithms in practice; the best theoretical bounds are reached when comparisons are done in a specific order; finally there exist some algorithms for which the order in which the comparisons are done is not relevant (such is the brute force algorithm).

1 From left to right

Hashing provides a simple method that avoids the quadratic number of character comparisons in most practical situations, and that runs in linear time under reasonable probabilistic assumptions. It has been introduced by Harrison and later fully analyzed by Karp and Rabin (see chapter 4).

Assuming that the pattern length is no greater than the memory-word size of the machine, the Shift Or algorithm (see chapter 5) is an efficient algorithm to solve the exact string matching problem and it adapts easily to a wide range of approximate string matching problems.

The first linear-time string matching algorithm is from Morris and Pratt (see chapter 6). It has been improved by Knuth, Morris, and Pratt (see chapter 7). The search behaves like a recognition process by automaton, and a character of the text is compared to a character of the pattern no more than $\log_\Phi(m+1)$ (Φ is the golden ratio $(1+\sqrt{5})/2$). Hancart proved that this delay of a related algorithm discovered by Simon makes no more than $1 + \log_2 m$ comparisons per text character (see chapter 8). Those three algorithms perform at most $2n - 1$ text character comparisons in the worst case.

The search with a Deterministic Finite Automaton performs exactly n text character inspections but it requires an extra space in $O(m \times \sigma)$ (see chapter 3). The Forward Dawg Matching algorithm performs exactly the same number of text character inspections using the suffix automaton of the pattern (see chapter 13).

The Apostolico-Crochemore algorithm (see chapter 11) is a simple algorithm which performs $\frac{3}{2}n$ text character comparisons in the worst case.

The Not So Naive algorithm (see chapter 12) is a very simple algorithm with a quadratic worst case time complexity but it requires a preprocessing phase in constant time and space and is slightly sub-linear in the average case.

2 From right to left

The Boyer-Moore algorithm (see chapter 14) is considered as the most efficient string matching algorithm in usual applications. A simplified version of it (or the entire algorithm) is often implemented in text editors for the "search" and "substitute" commands. Cole proved that the maximum number of character comparisons is tightly bounded by $3n$ after the preprocessing for non-periodic patterns. It has a quadratic worst case time for periodic patterns.

Several variants of the Boyer-Moore algorithm avoid its quadratic behavior. The most efficient solutions in term of number of character comparisons have been designed by Galil (see chapter 15), Smyth (see chapter 16), Crochemore *et alii* (Turbo-BM) (see chapter 17), Apostolico and Giancarlo (see chapter 18), and Colussi (Reverse Colussi) (see chapter 19). Empirical results show that variations of the Boyer-Moore algorithm and algorithms based on the suffix automaton by Crochemore *et alii* (Reverse Factor and Turbo Reverse Factor) (see chapters 29 and 30) or the suffix oracle by Allauzen *et alii* (Backward Suffix Oracle Matching) (see chapter 31) are the most efficient in practice. The Backward Nondeterministic Dawg Matching algorithm (see chapter 32) of Navarro and Raffinot is an implementation of the Reverse Factor algorithm using bitwise techniques.

The Zhu-Takaoka (see chapter 25) and Berry-Ravindran (see chapter 26) algorithms are variants of the Boyer-Moore algorithm which require an extra space in $O(\sigma^2)$.

The Fast Search algorithm (see chapter 21) is an hybrid between the Boyer-Moore algorithm and the Horspool algorithm.

3 In a specific order

The two first linear optimal space string matching algorithms are due to Galil-Seiferas (see chapter 33) and Crochemore-Perrin (Two Way) (see chapter 34). They partition the pattern in two parts, they first search for the right part of the pattern from left to right and then if no mismatch occurs they search for the left part.

The algorithms of Colussi (see chapter 9) and Galil-Giancarlo (see chapter 10) partition the set of pattern positions into two subsets. They first search for the pattern characters which positions are in the first subset from left to right and then if no mismatch occurs they search for the remaining characters from left to right. The Colussi algorithm is an improvement over the Knuth-Morris-Pratt algorithm and performs at most $\frac{3}{2}n$ text character comparisons in the worst case. The Galil-Giancarlo algorithm improves the Colussi algorithm in one special case which enables it to perform at most $\frac{4}{3}n$ text character comparisons in the worst case.

Sunday's Optimal Mismatch (see chapter 36) and Maximal Shift (see chapter 37) algorithms sort the pattern positions according their character frequency and their leading shift respectively.

Skip Search (see chapter 38), KMPSkip Search (see chapter 39) and Alpha Skip Search (see chapter 40) algorithms by Charras and Lecroq use buckets to determine starting positions on the pattern in the text.

4 In any order

The Horspool algorithm (see chapter 20) is a variant of the Boyer-Moore algorithm, it uses only one of its shift functions and the order in which the text character comparisons are performed is irrelevant. This is also true for other variants such as the Quick Search algorithm (see chapter 22) of Sunday, the Turbo Search algorithm (see chapter 23) of Tamm, the Tuned Boyer-Moore algorithm (see chapter 24) of Hume and Sunday, the Smith algorithm (see chapter 27) and the Raita algorithm (see chapter 28).

5 Conventions

We will consider practical searches. We will assume that the alphabet is the set of ASCII codes or any subset of it. The algorithms are presented in C programming language, thus a word w of length ℓ can be written $w[0\,.\,.\,\ell-1]$; the characters are $w[0], \ldots, w[\ell-1]$ and $w[\ell]$ contains the special end character (null character) that cannot occur anywhere within any word but in the end. Both words the pattern and the text reside in main memory.

Let us introduce some definitions.

5.1 Definitions

A word u is a **prefix** of a word w is there exists a word v (possibly empty) such that $w = uv$.

A word v is a **suffix** of a word w is there exists a word u (possibly empty) such that $w = uv$.

A word z is a **substring** or a **subword** or a **factor** of a word w is there exist two words u and v (possibly empty) such that $w = uzv$.

A integer p is a **period** of a word w if for i, $0 \le i < m - p$, $w[i] = w[i+p]$. The smallest period of w is called **the period** of w, it is denoted by $per(w)$. A word w of length ℓ is **periodic** if the length of its smallest period is smaller or equal to $\ell/2$, otherwise it is **non-periodic**.

A word w is **basic** if it cannot be written as a power of another word: there exist no word z and no integer k such that $w = z^k$.

A word z is a **border** of a word w if there exist two words u and v such that $w = uz = zv$, z is both a prefix and a suffix of w. Note that in this case $|u| = |v|$ is a period of w.

The **reverse** of a word w of length ℓ denoted by w^R is the mirror image of w: $w^R = w[\ell-1]w[\ell-2]\ldots w[1]w[0]$.

A Deterministic Finite Automaton (DFA) $\mathcal{A}$ is a quadruple (Q, q_0, T, E) where:

- Q is a finite set of states;
- $q_0 \in Q$ is the initial state;
- $T \subseteq Q$ is the set of terminal states;
- $E \subseteq (Q \times \Sigma \times Q)$ is the set of transitions.

The language $\mathcal{L}(\mathcal{A})$ defined by $\mathcal{A}$ is the following set:

$$\{w \in \Sigma^* \mid \exists q_0, \ldots, q_n, n = |w|, q_n \in T, \forall 0 \le i < n, (q_i, w[i], q_{i+1}) \in E\}$$

For each exact string matching algorithm presented in the present book we first give its main features, then we explained how it works before giving its C code. After that we show its behavior on a typical example where $x =$ GCAGAGAG of length 8 and $y =$ GCATCGCAGAGAGTATACAGTACG of length 24. At each attempt, matches are materialized in light gray while mismatches are shown in dark gray. A number indicates the order in which the character comparisons are done except for the algorithms using automata where the number represents the state reached after the character inspection. Finally we give a list of references where the reader will find more detailed presentations and proofs of the algorithm.

5.2 Implementations

Code optimization is beyond the scope of this book thus the programs that are presented in the next chapters are very close to the algorithms given by their designers.

Characters are implemented by **unsigned char**: for that we define a type **Character** by

```
typedef unsigned char Character
```

and strings are implemented by **unsigned char ***: for that we define a type `String` by

```
typedef unsigned char * String.
```

The strings x and y are implemented by two variables x and y, respectively, of type `String`. The variable x has size at least $m+1$: the first m elements contain the string x and the $(m+1)$-th element contains the special end character (null character) that cannot occur anywhere within any word but in the end. The variable y has size at least $m+n$: the first n elements contain the string y and the $(n+1)$-th element contains the special end character (null character) that cannot occur anywhere within any word but in the end.

In all the functions:

- i is an index in the pattern x;

- j is an index in the text y;

- ASIZE is the size of the alphabet;

- XSIZE is the maximum pattern length;

- YSIZE is the maximum text length;

- OUTPUT(j) is used to output the left position j of an occurrence of x in y: it can be replaced, for instance, by `printf("%d\n", j)`.

In this book, we will use classical tools. One of them is a linked list of integer. It will be defined in C as follows:

```
struct _cell {
   int element;
   struct _cell *next;
};

typedef struct _cell *List;
```

Other important structures are tries and automata, specifically suffix automata (see chapter 29). Basically automata are directed graphs. We will use the following interface to manipulate automata (assuming that vertices will be associated with positive integers):

```c
/* returns a new data structure for
   a graph with v vertices and e edges */
Graph newGraph(int v, int e);

/* returns a new data structure for
   a automaton with v vertices and e edges */
Graph newAutomaton(int v, int e);

/* returns a new data structure for
   a suffix automaton with v vertices and e edges */
Graph newSuffixAutomaton(int v, int e);

/* returns a new data structure for
   a trie with v vertices and e edges */
Graph newTrie(int v, int e);

/* returns a new vertex for graph g */
int newVertex(Graph g);

/* returns the initial vertex of graph g */
int getInitial(Graph g);

/* returns true if vertex v is terminal in graph g */
boolean isTerminal(Graph g, int v);

/* set vertex v to be terminal in graph g */
void setTerminal(Graph g, int v);

/* returns the target of edge from vertex v
   labeled by character c in graph g */
int getTarget(Graph g, int v, Character c);

/* add the edge from vertex v to vertex t
   labeled by character c in graph g */
void setTarget(Graph g, int v, Character c, int t);

/* returns the suffix link of vertex v in graph g */
int getSuffixLink(Graph g, int v);
```

```
/* set the suffix link of vertex v
   to vertex s in graph g */
void setSuffixLink(Graph g, int v, int s);

/* returns the length of vertex v in graph g */
int getLength(Graph g, int v);

/* set the length of vertex v to integer ell in graph g */
void setLength(Graph g, int v, int ell);

/* returns the position of vertex v in graph g */
int getPosition(Graph g, int v);

/* set the length of vertex v to integer ell in graph g */
void setPosition(Graph g, int v, int p);

/* returns the shift of the edge from vertex v
   labeled by character c in graph g */
int getShift(Graph g, int v, Character c);

/* set the shift of the edge from vertex v
   labeled by character c to integer s in graph g */
void setShift(Graph g, int v, Character c, int s);

/* copies all the characteristics of vertex source
   to vertex target in graph g */
void copyVertex(Graph g, int target, int source);
```

A possible implementation is given in appendix I.

CHAPTER 2

BRUTE FORCE ALGORITHM

1 Main features

- no preprocessing phase;

- constant extra space needed;

- always shifts the window by exactly 1 position to the right;

- comparisons can be done in any order;

- searching phase in $O(m \times n)$ time complexity;

- $2n$ expected text character comparisons.

2 Description

The brute force algorithm consists in checking, at all positions in the text between 0 and $n - m$, whether an occurrence of the pattern starts there or not. Then, after each attempt, it shifts the pattern by exactly one position to the right.

The brute force algorithm requires no preprocessing phase, and a constant extra space in addition to the pattern and the text. During the searching phase the text character comparisons can be done in any order. The time complexity of this searching phase is $O(m \times n)$ (when searching for $\mathtt{a}^{m-1}\mathtt{b}$ in $\mathtt{a}^n$ for instance). The expected number of text character comparisons is $2n$.

3 The C code

```
void BF(String x, int m, String y, int n) {
   int i, j;

   /* Searching */
   for (j = 0; j <= n - m; ++j) {
      for (i = 0; i < m && x[i] == y[i + j]; ++i);
      if (i >= m)
         OUTPUT(j);
   }
}
```

This algorithm can be rewriting to give a more efficient algorithm in practice as follows:

```
#define EOS '\0'

void BF(String x, int m, String y, int n) {
   String yb;

   /* Searching */
   for (yb = y; *y != EOS; ++y)
      if (memcmp(x, y, m) == 0)
         OUTPUT(y - yb);
}
```

However code optimization is beyond the scope of this book.

4 The example
Searching phase

First attempt:

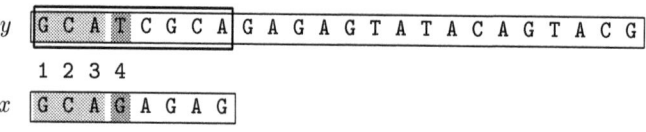

Shift by 1

Second attempt:

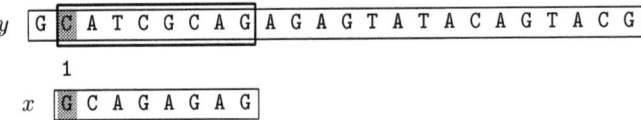

Shift by 1

Third attempt:

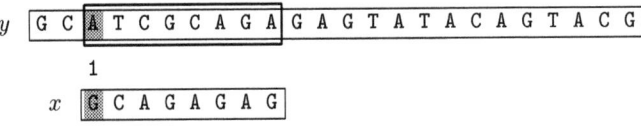

Shift by 1

Fourth attempt:

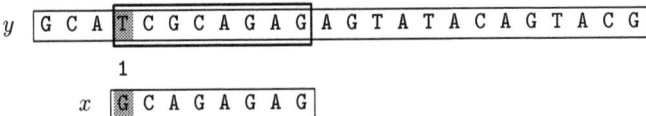

Shift by 1

Fifth attempt:

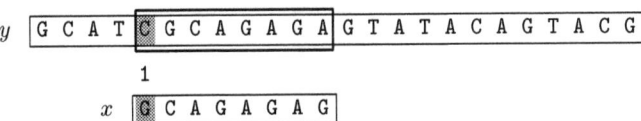

Shift by 1

Sixth attempt:

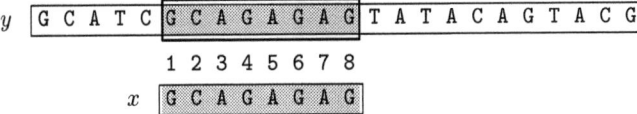

Shift by 1

12   *Handbook of Exact String Matching Algorithms*

Seventh attempt:

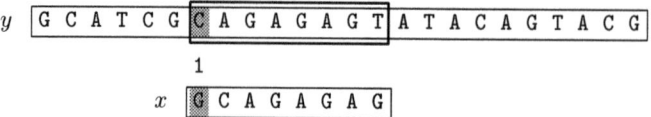

Shift by 1

Eighth attempt:

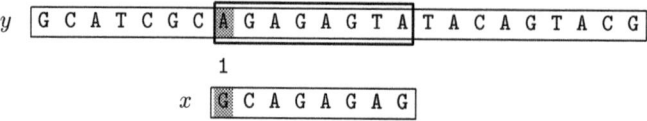

Shift by 1

Ninth attempt:

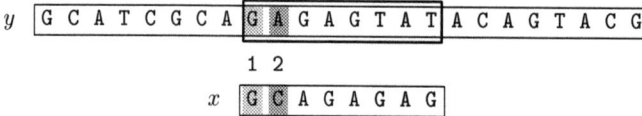

Shift by 1

Tenth attempt:

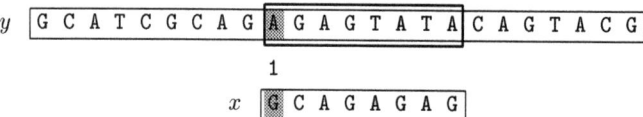

Shift by 1

Eleventh attempt:

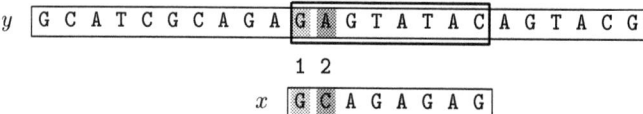

Shift by 1

2. BRUTE FORCE ALGORITHM

Twelfth attempt:

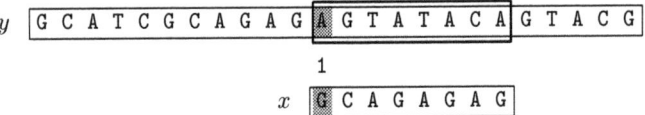

Shift by 1

Thirteenth attempt:

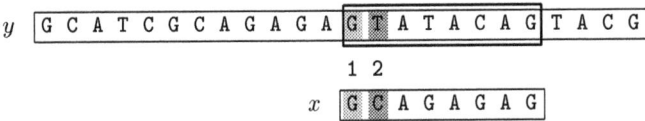

Shift by 1

Fourteenth attempt:

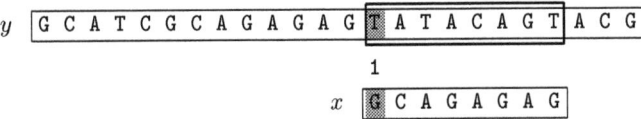

Shift by 1

Fifteenth attempt:

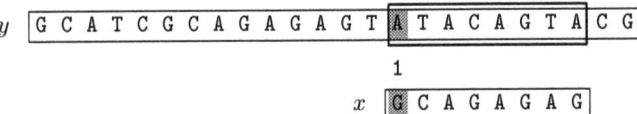

Shift by 1

Sixteenth attempt:

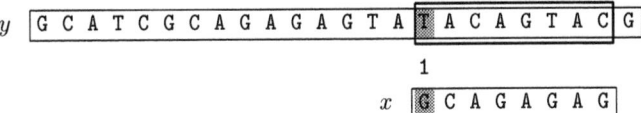

Shift by 1

Seventeenth attempt:

```
y  G C A T C G C A G A G A G T A T A C A G T A C G
                                  1
x                                 G C A G A G A G
```
Shift by 1

The brute force algorithm performs 30 text character comparisons on the example.

CHAPTER 3

SEARCH WITH AN AUTOMATON

1 Main features

- builds the minimal Deterministic Finite Automaton recognizing the language Σ^*x;
- extra space in $O(m \times \sigma)$ if the automaton is stored in a direct access table;
- preprocessing phase in $O(m \times \sigma)$ time complexity;
- searching phase in $O(n)$ time complexity if the automaton is stored in a direct access table, $O(n \times \log \sigma)$ otherwise.

2 Description

Searching a word x with an automaton consists first in building the minimal Deterministic Finite Automaton (DFA) $\mathcal{A}(x)$ recognizing the language Σ^*x.

The DFA $\mathcal{A}(x) = (Q, q_0, T, E)$ recognizing the language Σ^*x is defined as follows:

- Q is the set of all the prefixes of x:
  $$Q = \{\varepsilon, x[0], x[0\mathinner{.\,.}1], \ldots, x[0\mathinner{.\,.}m-2], x\},$$
- $q_0 = \varepsilon$,
- $T = \{x\}$,
- for $q \in Q$ (q is a prefix of x) and $a \in \Sigma$, $(q, a, qa) \in E$ if and only if qa is also a prefix of x, otherwise $(q, a, p) \in E$ such that p is the longest suffix of qa which is a prefix of x.

The DFA $\mathcal{A}(x)$ can be constructed in $O(m+\sigma)$ time and $O(m \times \sigma)$ space.

Once the DFA $\mathcal{A}(x)$ is build, searching for a word x in a text y consists in parsing the text y with the DFA $\mathcal{A}(x)$ beginning with the initial state q_0. Each time the terminal state is encountered an occurrence of x is reported.

The searching phase can be performed in $O(n)$ time if the automaton is stored in a direct access table, in $O(n \times \log \sigma)$ otherwise.

3 The C code

```
void preAut(String x, int m, Graph aut) {
   int i, state, target, oldTarget;

   for (state = getInitial(aut), i = 0; i < m; ++i) {
      oldTarget = getTarget(aut, state, x[i]);
      target = newVertex(aut);
      setTarget(aut, state, x[i], target);
      copyVertex(aut, target, oldTarget);
      state = target;
   }
   setTerminal(aut, state);
}


void AUT(String x, int m, String y, int n) {
   int j, state;
   Graph aut;

   /* Preprocessing */
   aut = newAutomaton(m + 1, (m + 1)*ASIZE);
   preAut(x, m, aut);

   /* Searching */
   for (state = getInitial(aut), j = 0; j < n; ++j) {
      state = getTarget(aut, state, y[j]);
      if (isTerminal(aut, state))
         OUTPUT(j - m + 1);
   }
}
```

4 The example

$\Sigma = \{\text{A}, \text{C}, \text{G}, \text{T}\}$

$Q = \{\varepsilon, \text{G}, \text{GC}, \text{GCA}, \text{GCAG}, \text{GCAGA}, \text{GCAGAG}, \text{GCAGAGA}, \text{GCAGAGAG}\}$

$q_0 = \varepsilon$

$T = \{\text{GCAGAGAG}\}$

3. SEARCH WITH AN AUTOMATON

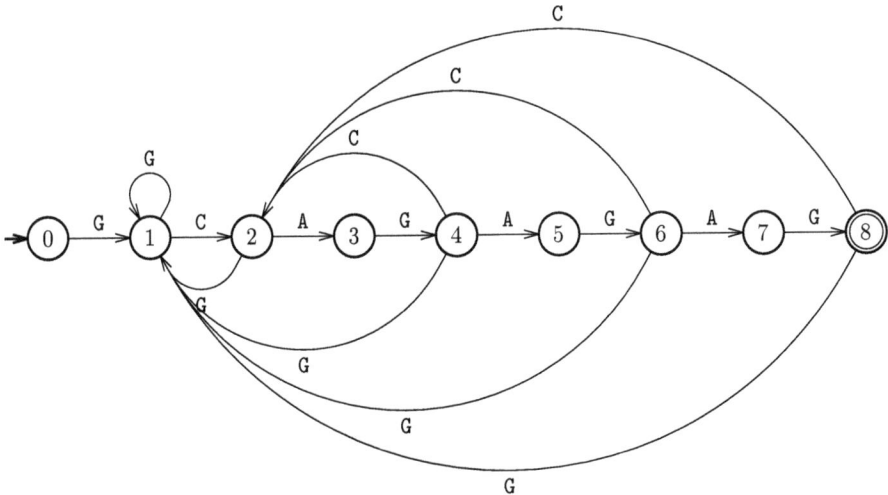

The states are labeled by the length of the prefix they are associated with.
Missing transitions are leading to the initial state 0.

Searching phase

The initial state is 0.

y | **G** C A T C G C A G A G A G T A T A C A G T A C G |
    1

y | G **C** A T C G C A G A G A G T A T A C A G T A C G |
      2

y | G C **A** T C G C A G A G A G T A T A C A G T A C G |
        3

y | G C A **T** C G C A G A G A G T A T A C A G T A C G |
          0

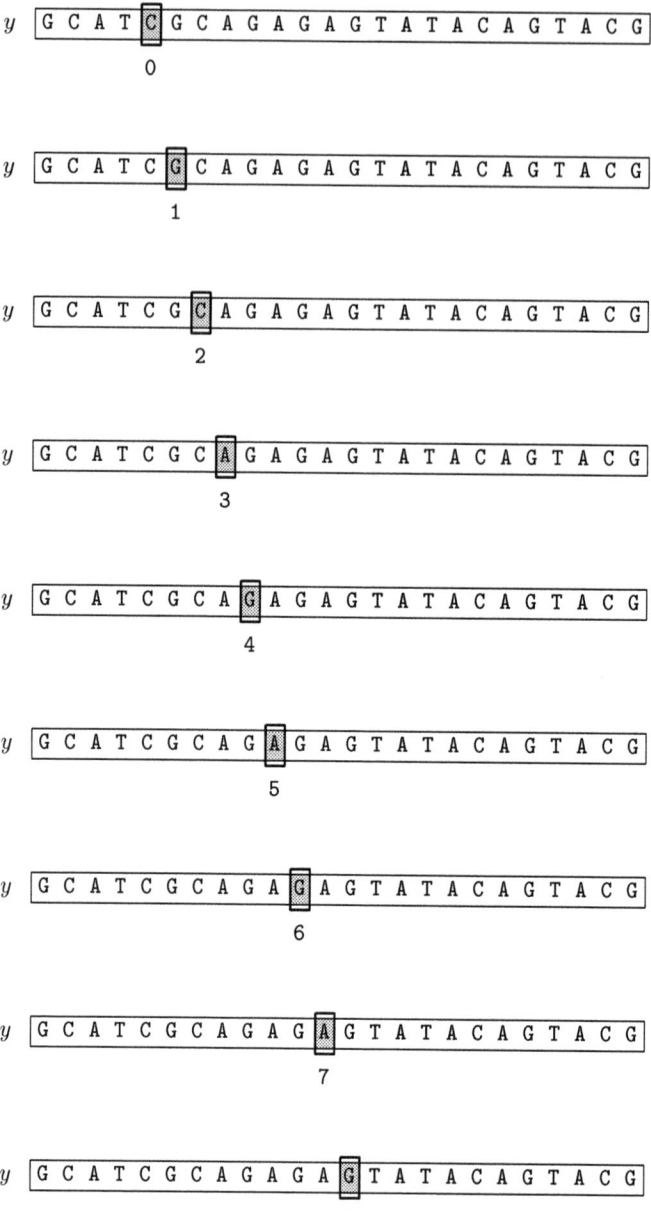

The state 8 is terminal, thus an occurrence of the pattern has been found.

3. SEARCH WITH AN AUTOMATON

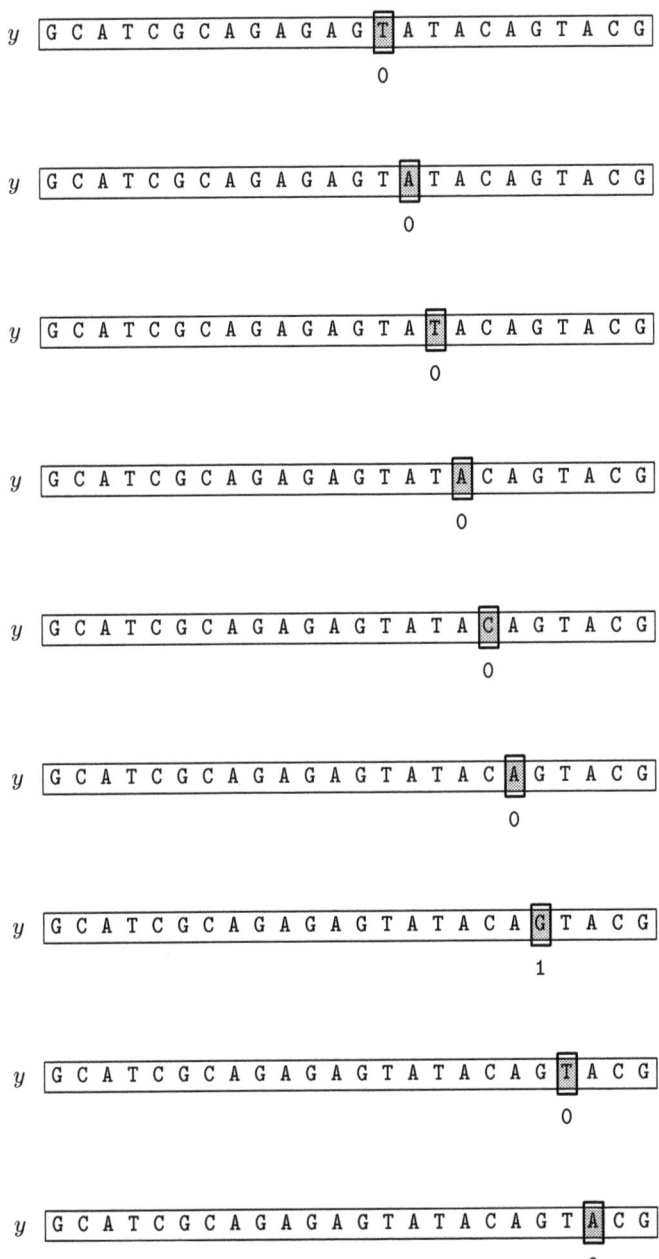

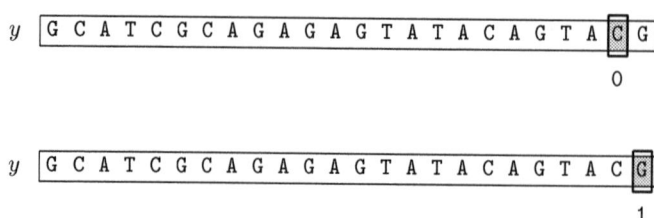

The search by automaton performs exactly 24 text character inspections on the example.

5 References

- CORMEN, T.H., LEISERSON, C.E., RIVEST, R.L., 1990, *Introduction to Algorithms*, Chapter 34, pp 853–885, MIT Press.

- CROCHEMORE, M., 1997, Off-line serial exact string searching, in *Pattern Matching Algorithms*, A. Apostolico and Z. Galil eds., Chapter 1, pp 1–53, Oxford University Press.

- CROCHEMORE, M., HANCART, C., 1997, Automata for Matching Patterns, in *Handbook of Formal Languages*, Volume 2, Linear Modeling: Background and Application, G. Rozenberg and A. Salomaa eds., Chapter 9, pp 399–462, Springer-Verlag, Berlin.

- CROCHEMORE, M., HANCART, C., LECROQ, T., 2001, *Algorithmique du texte*, Vuibert.

- GONNET, G.H., BAEZA-YATES, R.A., 1991, *Handbook of Algorithms and Data Structures in Pascal and C*, 2nd Edition, Chapter 7, pp. 251–288, Addison-Wesley Publishing Company.

- HANCART, C., 1993, *Analyse exacte et en moyenne d'algorithmes de recherche d'un motif dans un texte*, Thèse de doctorat de l'Université de Paris 7, France.

CHAPTER 4

KARP-RABIN ALGORITHM

1 Main features

- uses an hashing function;
- preprocessing phase in $O(m)$ time complexity and constant space;
- searching phase in $O(m \times n)$ time complexity;
- $O(m+n)$ expected running time.

2 Description

Hashing provides a simple method to avoid a quadratic number of character comparisons in most practical situations. Instead of checking at each position of the text if the pattern occurs, it seems to be more efficient to check only if the contents of the window "looks like" the pattern. In order to check the resemblance between these two words an hashing function is used. To be helpful for the string matching problem an hashing function $hash$ should have the following properties:

- efficiently computable;
- highly discriminating for strings;
- $hash(y[j+1 \mathinner{.\,.} j+m])$ must be easily computable from $hash(y[j \mathinner{.\,.} j+m-1])$ and $y[j+m]$:

$$hash(y[j+1 \mathinner{.\,.} j+m]) = rehash(y[j], y[j+m], hash(y[j \mathinner{.\,.} j+m-1])).$$

For a word w of length m let $hash(w)$ be defined as follows:

$$hash(w[0 \mathinner{.\,.} m-1]) = (w[0] \times 2^{m-1} + w[1] \times 2^{m-2} + \cdots + w[m-1] \times 2^0) \bmod q$$

where q is a large number. Then, for $a, b \in \Sigma$ and h a number

$$rehash(a, b, h) = ((h - a \times 2^{m-1}) \times 2 + b) \bmod q.$$

The preprocessing phase of the Karp-Rabin algorithm consists in computing $hash(x)$ and $hash(y[0..m-1])$. It can be done in constant space and $O(m)$ time.

During the searching phase, it is enough to compare the value $hash(x)$ with $hash(y[j..j+m-1])$ for $0 \leq j \leq n-m$. If an equality is found, it is still necessary to check the equality $x = y[j..j+m-1]$ character by character.

The time complexity of the searching phase of the Karp-Rabin algorithm is $O(m \times n)$ (when searching for a^m in a^n for instance). Its expected number of text character comparisons is $O(m+n)$.

3 The C code

In the following function KR all the multiplications by 2 are implemented by shifts. Furthermore, the computation of the modulus function is avoided by using the implicit modular arithmetic given by the hardware that forgets carries in integer operations. So, q is chosen as the maximum value for an integer.

```
#define REHASH(a, b, h) ((((h) - (a)*d) << 1) + (b))

void KR(String x, int m, String y, int n) {
   int d, hx, hy, i, j;

   /* Preprocessing */
   /* computes d = 2^(m-1) with the left-shift operator */
   for (d = i = 1; i < m; ++i)
      d = (d<<1);
   for (hy = hx = i = 0; i < m; ++i) {
      hx = ((hx<<1) + x[i]);
      hy = ((hy<<1) + y[i]);
   }

   /* Searching */
   j = 0;
   while (j <= n-m) {
      if (hx == hy && memcmp(x, y + j, m) == 0)
         OUTPUT(j);
      hy = REHASH(y[j], y[j + m], hy);
      ++j;
   }
}
```

4 The example

$hash(\text{GCAGAGAG}) = 17597$

Searching phase

First attempt:

| y | G C A T C G C A | G A G A G T A T A C A G T A C G |

| x | G C A G A G A G |

$hash(y[0..7]) = 17819$

Second attempt:

| y | G | C A T C G C A G | A G A G T A T A C A G T A C G |

| x | G C A G A G A G |

$hash(y[1..8]) = 17533$

Third attempt:

| y | G C | A T C G C A G A | G A G T A T A C A G T A C G |

| x | G C A G A G A G |

$hash(y[2..9]) = 17979$

Fourth attempt:

| y | G C A | T C G C A G A G | A G T A T A C A G T A C G |

| x | G C A G A G A G |

$hash(y[3..10]) = 19389$

Fifth attempt:

| y | G C A T | C G C A G A G A | G T A T A C A G T A C G |

| x | G C A G A G A G |

$hash(y[4..11]) = 17339$

Sixth attempt:

```
y  G C A T C G C A G A G A G T A T A C A G T A C G
             1 2 3 4 5 6 7 8
         x   G C A G A G A G
```

$hash(y[5..12]) = 17597 = hash(x)$ thus 8 character comparisons are necessary to be sure that an occurrence of the pattern has been found.

Seventh attempt:

```
y  G C A T C G C A G A G A G T A T A C A G T A C G
           x   G C A G A G A G
```

$hash(y[6..13]) = 17102$

Eighth attempt:

```
y  G C A T C G C A G A G A G T A T A C A G T A C G
             x   G C A G A G A G
```

$hash(y[7..14]) = 17117$

Ninth attempt:

```
y  G C A T C G C A G A G A G T A T A C A G T A C G
               x   G C A G A G A G
```

$hash(y[8..15]) = 17678$

Tenth attempt:

```
y  G C A T C G C A G A G A G T A T A C A G T A C G
                 x   G C A G A G A G
```

$hash(y[9..16]) = 17245$

Eleventh attempt:

```
y  G C A T C G C A G A G A G T A T A C A G T A C G
                   x   G C A G A G A G
```

$hash(y[10..17]) = 17917$

4. KARP-RABIN ALGORITHM

Twelfth attempt:

y | G C A T C G C A G A G | A G T A T A C A | G T A C G |

x | G C A G A G A G |

$hash(y[11..18]) = 17723$

Thirteenth attempt:

y | G C A T C G C A G A G A | G T A T A C A G | T A C G |

x | G C A G A G A G |

$hash(y[12..19]) = 18877$

Fourteenth attempt:

y | G C A T C G C A G A G A G | T A T A C A G T | A C G |

x | G C A G A G A G |

$hash(y[13..20]) = 19662$

Fifteenth attempt:

y | G C A T C G C A G A G A G T | A T A C A G T A | C G |

x | G C A G A G A G |

$hash(y[14..21]) = 17885$

Sixteenth attempt:

y | G C A T C G C A G A G A G T A | T A C A G T A C | G |

x | G C A G A G A G |

$hash(y[15..22]) = 19197$

Seventeenth attempt:

y | G C A T C G C A G A G A G T A T | A C A G T A C G |

x | G C A G A G A G |

$hash(y[16..23]) = 16961$

The Karp-Rabin algorithm performs 17 comparisons on hashing values and 8 character comparisons on the example.

5 References

- AHO, A.V., 1990, Algorithms for Finding Patterns in Strings, in *Handbook of Theoretical Computer Science, Volume A, Algorithms and complexity*, J. van Leeuwen ed., Chapter 5, pp 255–300, Elsevier, Amsterdam.

- CORMEN, T.H., LEISERSON, C.E., RIVEST, R.L., 1990, *Introduction to Algorithms*, Chapter 34, pp 853–885, MIT Press.

- CROCHEMORE, M., HANCART, C., 1999, Pattern Matching in Strings, in *Algorithms and Theory of Computation Handbook*, M.J. Atallah ed., Chapter 11, pp 11-1–11-28, CRC Press Inc., Boca Raton, FL.

- CROCHEMORE, M., RYTTER, W., 2002, *Jewels of Stringology*, World Scientific Press.

- GONNET, G.H., BAEZA-YATES, R.A., 1991, *Handbook of Algorithms and Data Structures in Pascal and C*, 2nd Edition, Chapter 7, pp. 251–288, Addison-Wesley Publishing Company.

- HANCART, C., 1993, *Analyse exacte et en moyenne d'algorithmes de recherche d'un motif dans un texte*, Thèse de doctorat de l'Université de Paris 7, France.

- CROCHEMORE, M., LECROQ, T., 1996, Pattern matching and text compression algorithms, in *CRC Computer Science and Engineering Handbook*, A.B. Tucker Jr ed., Chapter 8, pp 162–202, CRC Press Inc., Boca Raton, FL.

- KARP, R.M., RABIN, M.O., 1987, Efficient randomized pattern-matching algorithms, *IBM Journal on Research Development* **31**(2): 249–260.

- SEDGEWICK, R., 1988, *Algorithms*, Chapter 19, pp. 277–292, Addison-Wesley Publishing Company.

- SEDGEWICK, R., 1992, *Algorithms in C*, Chapter 19, Addison-Wesley Publishing Company.

- SMYTH, W. F., 2003, *Computing Patterns in Strings*, Pearson Addison Wesley.

- STEPHEN, G.A., 1994, *String Searching Algorithms*, World Scientific.

CHAPTER 5

SHIFT OR ALGORITHM

1 Main features

- uses bitwise techniques;
- efficient if the pattern length is no greater than the memory-word size of the machine;
- preprocessing phase in $O(m + \sigma)$ time and space complexity;
- searching phase in $O(n)$ time complexity (independent from the alphabet size and the pattern length);
- adapts easily to approximate string matching.

2 Description

The Shift Or algorithm uses bitwise techniques. Let R be a bit array of size m. Vector R_j is the value of the array R after text character $y[j]$ has been processed (see figure 5.1). It contains informations about all matches of prefixes of x that end at position j in the text. For $0 \leq i \leq m - 1$:

$$R_j[i] = \begin{cases} 0 & \text{if } x[0..i] = y[j-i..j], \\ 1 & \text{otherwise}. \end{cases}$$

The vector R_{j+1} can be computed after R_j as follows. For each $R_j[i] = 0$:

$$R_{j+1}[i+1] = \begin{cases} 0 & \text{if } x[i+1] = y[j+1], \\ 1 & \text{otherwise}, \end{cases}$$

and

$$R_{j+1}[0] = \begin{cases} 0 & \text{if } x[0] = y[j+1], \\ 1 & \text{otherwise}. \end{cases}$$

If $R_{j+1}[m-1] = 0$ then a complete match can be reported.

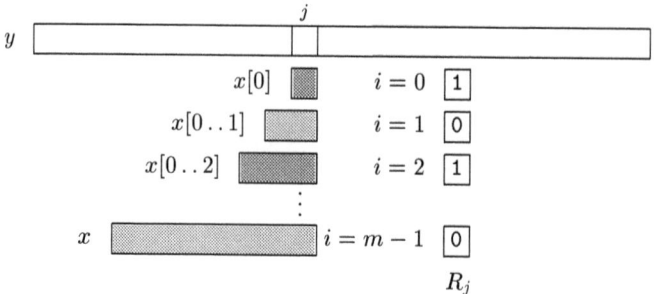

Figure 5.1. Meaning of vector R_j in the Shift-Or algorithm.

The transition from R_j to R_{j+1} can be computed very fast as follows. For each $c \in \Sigma$, let S_c be a bit array of size m such that:

for $0 \leq i \leq m-1$, $S_c[i] = 0$ if and only if $x[i] = c$.

The array S_c denotes the positions of the character c in the pattern x. Each S_c can be preprocessed before the search. And the computation of R_{j+1} reduces to two operations, shift and or:

$R_{j+1} = \text{Shift}(R_j)$ Or $S_{y[j+1]}$.

Assuming that the pattern length is no greater than the memory-word size of the machine, the space and time complexity of the preprocessing phase is $O(m + \sigma)$. The time complexity of the searching phase is $O(n)$, thus independent from the alphabet size and the pattern length. The Shift Or algorithm adapts easily to approximate string matching.

3 The C code

```
int preSo(String x, int m, unsigned int S[]) {
   unsigned int j, lim;
   int i;

   for (i = 0; i < ASIZE; ++i) S[i] = ~0;
   for (lim = i = 0, j = 1; i < m; ++i, j <<= 1) {
      S[x[i]] &= ~j;
      lim |= j;
   }
   lim = ~(lim>>1);
   return(lim);
}
```

```
void SO(String x, int m, String y, int n) {
   unsigned int lim, state;
   unsigned int S[ASIZE];
   int j;

   if (m > WORD)
      error("SO: Use pattern size <= word size");

   /* Preprocessing */
   lim = preSo(x, m, S);

   /* Searching */
   for (state = ~0, j = 0; j < n; ++j) {
      state = (state<<1) | S[y[j]];
      if (state < lim)
         OUTPUT(j - m + 1);
   }
}
```

4 The example

	S_A	S_C	S_G	S_T
G	1	1	0	1
C	1	0	1	1
A	0	1	1	1
G	1	1	0	1
A	0	1	1	1
G	1	1	0	1
A	0	1	1	1
G	1	1	0	1

		0	1	2	3	4	5	6	7	8	9	10	11	12	13	14	15	16	17	18	19	20	21	22	23
		G	C	A	T	C	G	C	A	G	A	G	A	G	T	A	T	A	C	A	G	T	A	C	G
0	G	0	1	1	1	1	0	1	1	0	1	0	1	0	1	1	1	1	1	1	0	1	1	1	0
1	C	1	0	1	1	0	1	0	1	1	1	1	1	1	1	1	1	1	1	1	1	1	1	1	1
2	A	1	1	0	1	1	1	1	0	1	1	1	1	1	1	1	1	1	1	1	1	1	1	1	1
3	G	1	1	1	1	1	1	1	1	0	1	1	1	1	1	1	1	1	1	1	1	1	1	1	1
4	A	1	1	1	1	1	1	1	1	1	0	1	1	1	1	1	1	1	1	1	1	1	1	1	1
5	G	1	1	1	1	1	1	1	1	1	1	0	1	1	1	1	1	1	1	1	1	1	1	1	1
6	A	1	1	1	1	1	1	1	1	1	1	1	0	1	1	1	1	1	1	1	1	1	1	1	1
7	G	1	1	1	1	1	1	1	1	1	1	1	1	0	1	1	1	1	1	1	1	1	1	1	1

As $R_{12}[7] = 0$ it means that an occurrence of x has been found at position $12 - 8 + 1 = 5$.

5 References

- BAEZA-YATES, R.A., GONNET, G.H., 1992, A new approach to text searching, *Communications of the ACM*. **35**(10):74–82.

- BAEZA-YATES, R.A., NAVARRO G., RIBEIRO-NETO B., 1999, Indexing and Searching, in *Modern Information Retrieval*, Chapter 8, pp 191–228, Addison-Wesley.

- CROCHEMORE, M., HANCART, C., LECROQ, T., 2001, *Algorithmique du texte*, Vuibert.

- CROCHEMORE, M., LECROQ, T., 1996, Pattern matching and text compression algorithms, in *CRC Computer Science and Engineering Handbook*, A.B. Tucker Jr ed., Chapter 8, pp 162–202, CRC Press Inc., Boca Raton, FL.

- GONNET, G.H., BAEZA-YATES, R.A., 1991, *Handbook of Algorithms and Data Structures in Pascal and C*, 2nd Edition, Chapter 7, pp. 251–288, Addison-Wesley Publishing Company.

- NAVARRO, G., RAFFINOT, M., 2002, *Flexible Pattern Matching in Strings Practical on-line search algorithms for texts and biological sequences*, Cambridge University Press.

- SMYTH, W. F., 2003, *Computing Patterns in Strings*, Pearson Addison Wesley.

- WU, S., MANBER, U., 1992, Fast text searching allowing errors, *Communications of the ACM*. **35**(10):83–91.

CHAPTER 6

MORRIS-PRATT ALGORITHM

1 Main Features

- performs the comparisons from left to right;
- preprocessing phase in $O(m)$ space and time complexity;
- searching phase in $O(n)$ time complexity (independent from the alphabet size);
- performs at most $2n-1$ text character comparisons during the searching phase;
- delay bounded by m.

2 Description

The design of the Morris-Pratt algorithm follows a tight analysis of the brute force algorithm (see chapter 2), and especially on the way this latter wastes the information gathered during the scan of the text.

Let us look more closely at the brute force algorithm. It is possible to improve the length of the shifts and simultaneously remember some portions of the text that match the pattern. This saves comparisons between characters of the pattern and characters of the text and consequently increases the speed of the search.

Consider an attempt at a left position j on y, that is when the window is positioned on the text factor $y[j\mathinner{.\,.}j+m-1]$. Assume that the first mismatch occurs between $x[i]$ and $y[i+j]$ with $0 < i < m$. Then, $x[0\mathinner{.\,.}i-1] = y[j\mathinner{.\,.}i+j-1] = u$ and $a = x[i] \neq y[i+j] = b$. When shifting, it is reasonable to expect that a prefix v of the pattern matches some suffix of the portion u of the text. The longest such prefix v is called the border of u (it occurs at both ends of u). This introduces the notation: let $mpNext[i]$ be the length of the longest border of $x[0\mathinner{.\,.}i-1]$ for $0 < i \leq m$. Then, after a shift, the comparisons can resume between characters $c = x[mpNext[i]]$ and $y[i+j] = b$ without missing any occurrence of x in y, and avoiding a backtrack on the text (see figure 6.1). The value of $mpNext[0]$ is set to

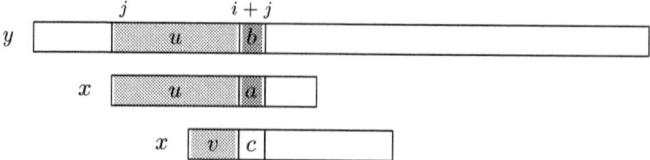

Figure 6.1. Shift in the Morris-Pratt algorithm: v is the border of u.

-1. The table *mpNext* can be computed in $O(m)$ space and time before the searching phase, applying the same searching algorithm to the pattern itself, as if $x = y$.

Then the searching phase can be done in $O(n)$ time. The Morris-Pratt algorithm performs at most $2n - 1$ text character comparisons during the searching phase. The delay (maximum number of comparisons for a single text character) is bounded by m.

3 The C code

```
void preMp(String x, int m, int mpNext[]) {
   int i, j;

   i = 0;
   j = mpNext[0] = -1;
   while (i < m) {
      while (j > -1 && x[i] != x[j])
         j = mpNext[j];
      mpNext[++i] = ++j;
   }
}


void MP(String x, int m, String y, int n) {
   int i, j, mpNext[XSIZE];

   /* Preprocessing */
   preMp(x, m, mpNext);

   /* Searching */
   i = j = 0;
   while (j < n) {
      while (i > -1 && x[i] != y[j])
```

```
        i = mpNext[i];
    i++;
    j++;
    if (i >= m) {
        OUTPUT(j - i);
        i = mpNext[i];
    }
   }
 }
}
```

4 The example

i	0	1	2	3	4	5	6	7	8
$x[i]$	G	C	A	G	A	G	A	G	
mpNext$[i]$	-1	0	0	0	1	0	1	0	1

Searching phase

First attempt:

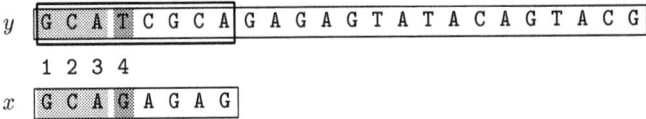

Shift by 3 ($i - $ mpNext$[i] = 3 - 0$)

Second attempt:

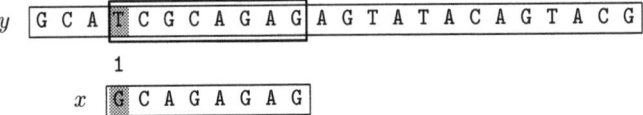

Shift by 1 ($i - $ mpNext$[i] = 0 - -1$)

Third attempt:

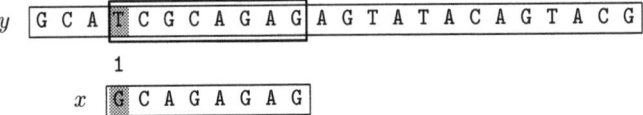

Shift by 1 ($i - $ mpNext$[i] = 0 - -1$)

Fourth attempt:

```
y  G C A T C G C A G A G A G T A T A C A G T A C G
             1 2 3 4 5 6 7 8
        x    G C A G A G A G
```
Shift by 7 ($i - mpNext[i] = 8 - 1$)

Fifth attempt:

```
y  G C A T C G C A G A G A G T A T A C A G T A C G
                                 1
                    x  G C A G A G A G
```
Shift by 1 ($i - mpNext[i] = 1 - 0$)

Sixth attempt:

```
y  G C A T C G C A G A G A G T A T A C A G T A C G
                                1
                      x  G C A G A G A G
```
Shift by 1 ($i - mpNext[i] = 0 - -1$)

Seventh attempt:

```
y  G C A T C G C A G A G A G T A T A C A G T A C G
                                  1
                        x  G C A G A G A G
```
Shift by 1 ($i - mpNext[i] = 0 - -1$)

Eighth attempt:

```
y  G C A T C G C A G A G A G T A T A C A G T A C G
                                    1
                          x  G C A G A G A G
```
Shift by 1 ($i - mpNext[i] = 0 - -1$)

Ninth attempt:

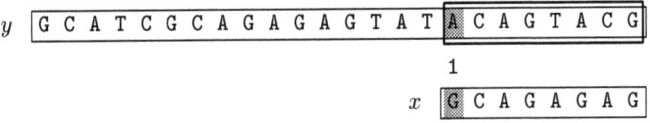

Shift by 1 $(i - mpNext[i] = 0 - -1)$

The Morris-Pratt algorithm performs 19 text character comparisons on the example.

5 References

- AHO, A.V., HOPCROFT, J.E., ULLMAN, J.D., 1974, *The design and analysis of computer algorithms*, 2nd Edition, Chapter 9, pp. 317–361, Addison-Wesley Publishing Company.

- BEAUQUIER, D., BERSTEL, J., CHRÉTIENNE, P., 1992, *Éléments d'algorithmique*, Chapter 10, pp 337–377, Masson, Paris.

- CROCHEMORE, M., 1997, Off-line serial exact string searching, in *Pattern Matching Algorithms*, A. Apostolico and Z. Galil eds., Chapter 1, pp 1–53, Oxford University Press.

- CROCHEMORE, M., HANCART, C., LECROQ, T., 2001, *Algorithmique du texte*, Vuibert.

- CROCHEMORE, M., RYTTER, W., 2002, *Jewels of Stringology*, World Scientific Press.

- HANCART, C., 1992, Une analyse en moyenne de l'algorithme de Morris et Pratt et de ses raffinements, in *Théorie des Automates et Applications, Actes des 2ᵉ Journées Franco-Belges*, D. Krob ed., Rouen, France, pp 99–110, PUR 176, Rouen, France.

- HANCART, C., 1993, *Analyse exacte et en moyenne d'algorithmes de recherche d'un motif dans un texte*, Thèse de doctorat de l'Université de Paris 7, France.

- MORRIS, JR, J.H., PRATT, V.R., 1970, *A linear pattern-matching algorithm*, Technical Report 40, University of California, Berkeley.

CHAPTER 7

KNUTH-MORRIS-PRATT ALGORITHM

1 Main Features

- performs the comparisons from left to right;
- preprocessing phase in $O(m)$ space and time complexity;
- searching phase in $O(n)$ time complexity (independent from the alphabet size);
- performs at most $2n-1$ text character comparisons during the searching phase;
- delay bounded by $\log_\Phi(m)$ where Φ is the golden ratio: $\Phi = \frac{1+\sqrt{5}}{2}$.

2 Description

The design of the Knuth-Morris-Pratt algorithm follows a tight analysis of the Morris-Pratt algorithm (see chapter 6). Let us look more closely at the Morris-Pratt algorithm. It is possible to improve the length of the shifts.

Consider an attempt at a left position j, that is when the window is positioned on the text factor $y[j \mathinner{.\,.} j+m-1]$. Assume that the first mismatch occurs between $x[i]$ and $y[i+j]$ with $0 < i < m$. Then, $x[0 \mathinner{.\,.} i-1] = y[j \mathinner{.\,.} i+j-1] = u$ and $a = x[i] \neq y[i+j] = b$. When shifting, it is reasonable to expect that a prefix v of the pattern matches some suffix of the portion u of the text. Moreover, if we want to avoid another immediate mismatch, the character following the prefix v in the pattern must be different from a. The longest such prefix v is called the tagged border of u (it occurs at both ends of u followed by different characters in x). This introduces the notation: let $kmpNext[i]$ be the length of the longest border of $x[0 \mathinner{.\,.} i-1]$ followed by a character c different from $x[i]$ and -1 if no such tagged border exits, for $0 < i \leq m$. Then, after a shift, the comparisons can resume between characters $x[kmpNext[i]]$ and $y[i+j]$ without missing any occurrence of x in y, and avoiding a backtrack on the text (see figure 7.1). The value of $kmpNext[0]$ is set to -1. The table $kmpNext$ can be computed in $O(m)$

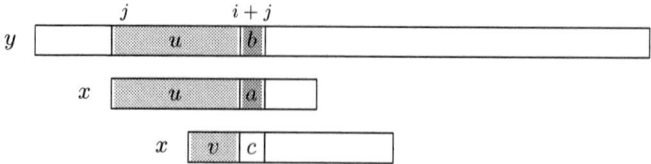

Figure 7.1. Shift in the Knuth-Morris-Pratt algorithm: v is a border of u and $a \neq c$.

space and time before the searching phase, applying the same searching algorithm to the pattern itself, as if $x = y$.

The searching phase can be performed in $O(n)$ time. The Knuth-Morris-Pratt algorithm performs at most $2n - 1$ text character comparisons during the searching phase. The delay (maximum number of comparisons for a single text character) is bounded by $\log_\Phi(m)$ where Φ is the golden ratio ($\Phi = \frac{1+\sqrt{5}}{2}$).

3 The C code

```
void preKmp(String x, int m, int kmpNext[]) {
   int i, j;

   i = 0;
   j = kmpNext[0] = -1;
   while (i < m) {
      while (j > -1 && x[i] != x[j])
         j = kmpNext[j];
      i++;
      j++;
      if (x[i] == x[j])
         kmpNext[i] = kmpNext[j];
      else
         kmpNext[i] = j;
   }
}
```

7. KNUTH-MORRIS-PRATT ALGORITHM

```
void KMP(String x, int m, String y, int n) {
   int i, j, kmpNext[XSIZE];

   /* Preprocessing */
   preKmp(x, m, kmpNext);

   /* Searching */
   i = j = 0;
   while (j < n) {
      while (i > -1 && x[i] != y[j])
         i = kmpNext[i];
      i++;
      j++;
      if (i >= m) {
         OUTPUT(j - i);
         i = kmpNext[i];
      }
   }
}
```

4 The example

i	0	1	2	3	4	5	6	7	8
$x[i]$	G	C	A	G	A	G	A	G	
$kmpNext[i]$	-1	0	0	-1	1	-1	1	-1	1

Searching phase

First attempt:

```
y  G C A T C G C A G A G A G T A T A C A G T A C G
   1 2 3 4
x  G C A G A G A G
```

Shift by 4 ($i - kmpNext[i] = 3 - -1$)

Second attempt:

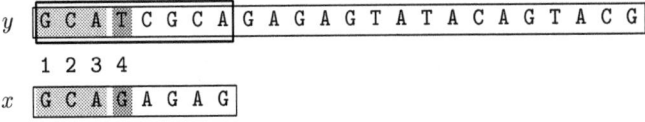

```
                 1
              x  G C A G A G A G
```

Shift by 1 ($i - kmpNext[i] = 0 - -1$)

Third attempt:

```
y  G C A T C G C A G A G A G T A T A C A G T A C G
             1 2 3 4 5 6 7 8
         x   G C A G A G A G
```

Shift by 7 ($i - kmpNext[i] = 8 - 1$)

Fourth attempt:

```
y  G C A T C G C A G A G A G T A T A C A G T A C G
                                 2
                             x   G C A G A G A G
```

Shift by 1 ($i - kmpNext[i] = 1 - 0$)

Fifth attempt:

```
y  G C A T C G C A G A G A G T A T A C A G T A C G
                                   1
                               x   G C A G A G A G
```

Shift by 1 ($i - kmpNext[i] = 0 - -1$)

Sixth attempt:

```
y  G C A T C G C A G A G A G T A T A C A G T A C G
                                     1
                                 x   G C A G A G A G
```

Shift by 1 ($i - kmpNext[i] = 0 - -1$)

Seventh attempt:

```
y  G C A T C G C A G A G A G T A T A C A G T A C G
                                       1
                                   x   G C A G A G A G
```

Shift by 1 ($i - kmpNext[i] = 0 - -1$)

Eighth attempt:

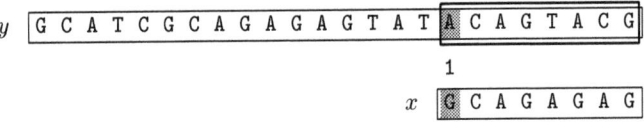

Shift by 1 ($i - \text{kmpNext}[i] = 0 - -1$)

The Knuth-Morris-Pratt algorithm performs 18 text character comparisons on the example.

5 References

- AHO, A.V., 1990, Algorithms for Finding Patterns in Strings, in *Handbook of Theoretical Computer Science, Volume A, Algorithms and complexity*, J. van Leeuwen ed., Chapter 5, pp 255–300, Elsevier, Amsterdam.

- AOE, J.-I., 1994, *Computer algorithms: string pattern matching strategies*, IEEE Computer Society Press.

- BAASE, S., VAN GELDER, A., 1999, *Computer Algorithms: Introduction to Design and Analysis*, 3rd Edition, Chapter 11, Addison-Wesley Publishing Company.

- BAEZA-YATES, R.A., NAVARRO G., RIBEIRO-NETO B., 1999, Indexing and Searching, in *Modern Information Retrieval*, Chapter 8, pp 191–228, Addison-Wesley.

- BEAUQUIER, D., BERSTEL, J., CHRÉTIENNE, P., 1992, *Éléments d'algorithmique*, Chapter 10, pp 337–377, Masson, Paris.

- CORMEN, T.H., LEISERSON, C.E., RIVEST, R.L., 1990, *Introduction to Algorithms*, Chapter 34, pp 853–885, MIT Press.

- CROCHEMORE, M., 1997, Off-line serial exact string searching, in *Pattern Matching Algorithms*, A. Apostolico and Z. Galil eds., Chapter 1, pp 1–53, Oxford University Press.

- CROCHEMORE, M., HANCART, C., 1999, Pattern Matching in Strings, in *Algorithms and Theory of Computation Handbook*, M.J. Atallah ed., Chapter 11, pp 11-1–11-28, CRC Press Inc., Boca Raton, FL.

- CROCHEMORE, M., HANCART, C., LECROQ, T., 2001, *Algorithmique du texte*, Vuibert.

- CROCHEMORE, M., LECROQ, T., 1996, Pattern matching and text compression algorithms, in *CRC Computer Science and Engineering Handbook*, A.B. Tucker Jr ed., Chapter 8, pp 162–202, CRC Press Inc., Boca Raton, FL.

- CROCHEMORE, M., RYTTER, W., 1994, *Text Algorithms*, Oxford University Press.

- CROCHEMORE, M., RYTTER, W., 2002, *Jewels of Stringology*, World Scientific Press.

- GONNET, G.H., BAEZA-YATES, R.A., 1991, *Handbook of Algorithms and Data Structures in Pascal and C*, 2nd Edition, Chapter 7, pp. 251–288, Addison-Wesley Publishing Company.

- GOODRICH, M.T., TAMASSIA, R., 1998, *Data Structures and Algorithms in JAVA*, Chapter 11, pp 441–467, John Wiley & Sons.

- GUSFIELD, D., 1997, *Algorithms on strings, trees, and sequences: Computer Science and Computational Biology*, Cambridge University Press.

- HANCART, C., 1992, Une analyse en moyenne de l'algorithme de Morris et Pratt et de ses raffinements, in *Théorie des Automates et Applications, Actes des 2ᵉ Journées Franco-Belges*, D. Krob ed., Rouen, France, pp 99–110, PUR 176, Rouen, France.

- HANCART, C., 1993, *Analyse exacte et en moyenne d'algorithmes de recherche d'un motif dans un texte*, Thèse de doctorat de l'Université de Paris 7, France.

- KNUTH, D.E., MORRIS, JR, J.H., PRATT, V.R., 1977, Fast pattern matching in strings, *SIAM Journal on Computing* **6**(1):323–350.

- NAVARRO, G., RAFFINOT, M., 2002, *Flexible Pattern Matching in Strings Practical on-line search algorithms for texts and biological sequences* , Cambridge University Press.

- SEDGEWICK, R., 1988, *Algorithms*, Chapter 19, pp. 277–292, Addison-Wesley Publishing Company.

- SEDGEWICK, R., 1992, *Algorithms in C*, Chapter 19, Addison-Wesley Publishing Company.

- SEDGEWICK, R., FLAJOLET, P., 1996, *An Introduction to the Analysis of Algorithms*, Chapter 7, Addison-Wesley Publishing Company.

- SMYTH, W. F., 2003, *Computing Patterns in Strings*, Pearson Addison Wesley.

- STEPHEN, G.A., 1994, *String Searching Algorithms*, World Scientific.

- WATSON, B.W., 1995, *Taxonomies and Toolkits of Regular Language Algorithms*, PhD Thesis, Eindhoven University of Technology, The Netherlands.

- WIRTH, N., 1986, *Algorithms & Data Structures*, Chapter 1, pp. 17–72, Prentice-Hall.

CHAPTER 8

SIMON ALGORITHM

1 Main features

- economical implementation of $\mathcal{A}(x)$ the minimal Deterministic Finite Automaton recognizing $\Sigma^* x$;

- preprocessing phase in $O(m)$ time and space complexity;

- searching phase in $O(n)$ time complexity (independent from the alphabet size);

- at most $2n - 1$ text character comparisons during the searching phase;

- delay bounded by $\min\{1 + \log_2 m, \sigma\}$.

2 Description

The main drawback of the search with the minimal DFA $\mathcal{A}(x)$ (see chapter 3) is the size of the automaton: $O(m \times \sigma)$. Simon noticed that there are only a few significant edges in $\mathcal{A}(x)$; they are:

- the forward edges going from the prefix of x of length k to the prefix of length $k+1$ for $0 \leq k < m$. There are exactly m such edges;

- the backward edges going from the prefix of x of length k to a smaller non-zero length prefix. The number of such edges is bounded by m.

The other edges are leading to the initial state and can then be deduced. Thus the number of significant edges is bounded by $2m$. Then for each state of the automaton it is only necessary to store the list of its significant outgoing edges.

Each state is represented by the length of its associated prefix minus 1 in order that each edge leading to state i, with $-1 \leq i \leq m - 1$ is labeled by $x[i]$ thus it is not necessary to store the labels of the edges. The forward edges can be easily deduced from the pattern, thus they are not stored. It only remains to store the significant backward edges.

We use a table L, of size $m - 1$, of linked lists. The element $L[i]$ gives the list of the targets of the edges starting from state i. In order to avoid to store the list for the state $m - 1$, during the computation of this table L, the integer ℓ is computed such that $\ell + 1$ is the length of the longest border of x.

The preprocessing phase of the Simon algorithm consists in computing the table L and the integer ℓ. It can be done in $O(m)$ space and time complexity.

The searching phase is analogous to the one of the search with an automaton. When an occurrence of the pattern is found, the current state is updated with the state ℓ. This phase can be performed in $O(n)$ time. The Simon algorithm performs at most $2n - 1$ text character comparisons during the searching phase. The delay (maximal number of comparisons for a single text character) is bounded by $\min\{1 + \log_2 m, \sigma\}$.

3   The C code

The description of a linked list List can be found section 5.

```c
int getTransition(String x, int m, int p, List L[],
                  Character c) {
   List cell;

   if (p < m - 1 && x[p + 1] == c)
      return(p + 1);
   else if (p > -1) {
      cell = L[p];
      while (cell != NULL)
         if (x[cell->element] == c)
            return(cell->element);
         else
            cell = cell->next;
      return(-1);
   }
   else
      return(-1);
}

void setTransition(int p, int q, List L[]) {
   List cell;

   cell = (List)malloc(sizeof(struct _cell));
   if (cell == NULL)
```

8. SIMON ALGORITHM

```
      error("SIMON/setTransition");
   cell->element = q;
   cell->next = L[p];
   L[p] = cell;
}


int preSimon(String x, int m, List L[]) {
   int i, k, ell;
   List cell;

   memset(L, NULL, (m - 2)*sizeof(List));
   ell = -1;
   for (i = 1; i < m; ++i) {
      k = ell;
      cell = (ell == -1 ? NULL : L[k]);
      ell = -1;
      if (x[i] == x[k + 1])
         ell = k + 1;
      else
         setTransition(i - 1, k + 1, L);
      while (cell != NULL) {
         k = cell->element;
         if (x[i] == x[k])
            ell = k;
         else
            setTransition(i - 1, k, L);
         cell = cell->next;
      }
   }
   return(ell);
}

void SIMON(String x, int m, String y, int n) {
   int j, ell, state;
   List L[XSIZE];

   /* Preprocessing */
   ell = preSimon(x, m, L);

   /* Searching */
   for (state = -1, j = 0; j < n; ++j) {
```

```
        state = getTransition(x, m, state, L, y[j]);
        if (state >= m - 1) {
            OUTPUT(j - m + 1);
            state = ell;
        }
    }
}
```

4 The example

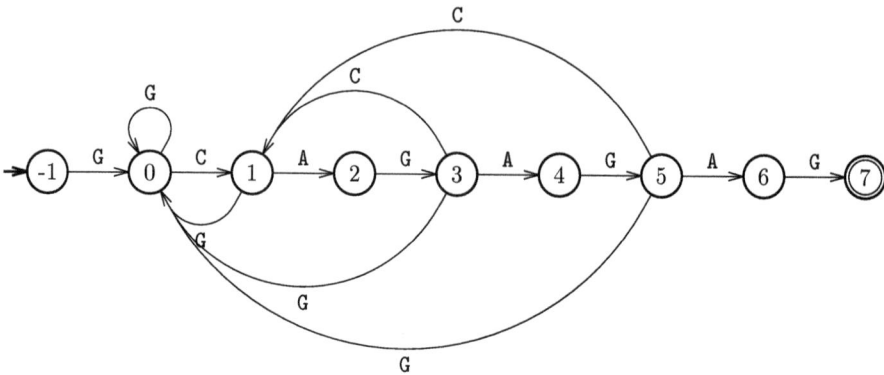

The states are labeled by the length of the prefix they are associated with minus 1.

i	0	1	2	3	4	5	6
$L[i]$	(0)	(0)	∅	(0,1)	∅	(0,1)	∅

Searching phase

The initial state is -1.

```
y  [G]C A T C G C A G A G A G T A T A C A G T A C G
    0
```

```
y   G[C]A T C G C A G A G A G T A T A C A G T A C G
     1
```

8. SIMON ALGORITHM

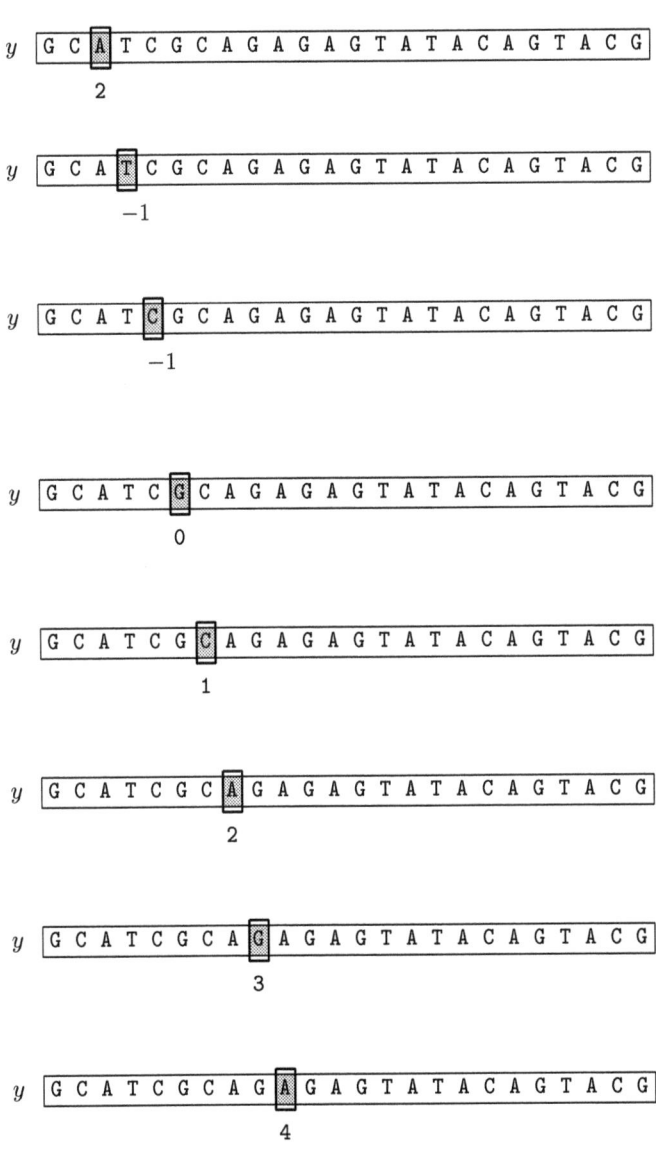

50 Handbook of Exact String Matching Algorithms

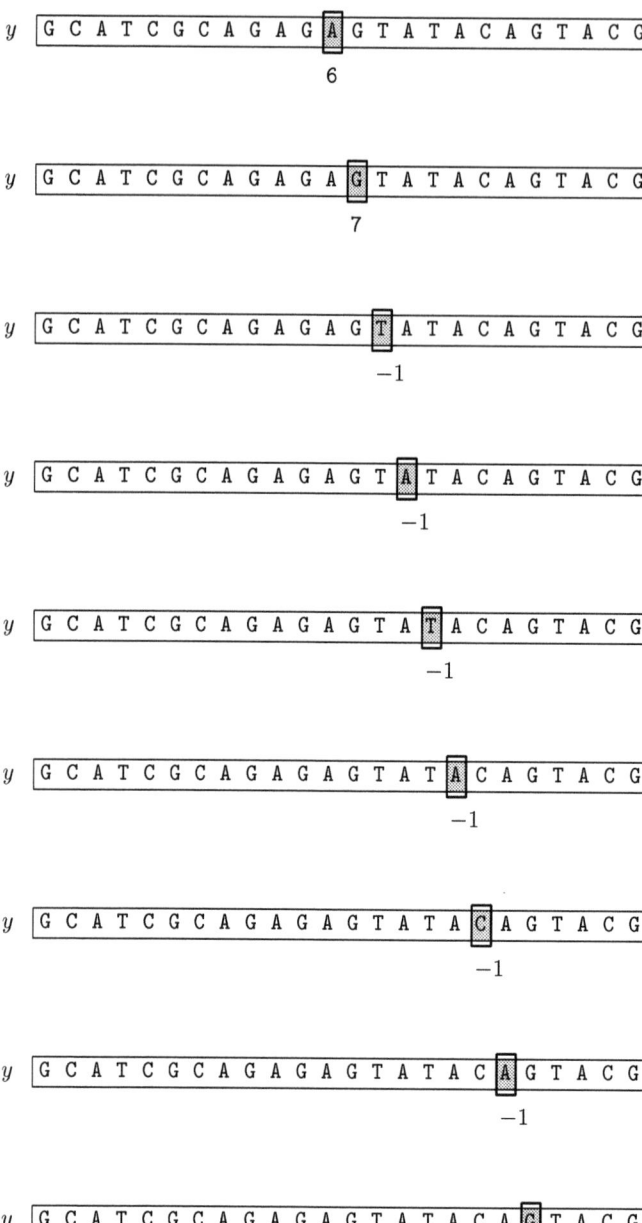

8. SIMON ALGORITHM 51

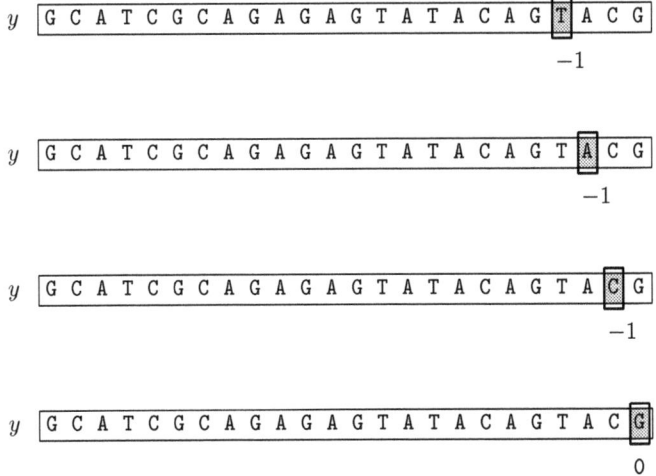

The Simon algorithm performs 24 text character comparisons on the example.

5 References

- BEAUQUIER, D., BERSTEL, J., CHRÉTIENNE, P., 1992, *Éléments d'algorithmique*, Chapter 10, pp 337–377, Masson, Paris.

- CROCHEMORE, M., 1997, Off-line serial exact string searching, in *Pattern Matching Algorithms*, A. Apostolico and Z. Galil eds., Chapter 1, pp 1–53, Oxford University Press.

- CROCHEMORE, M., HANCART, C., 1997, Automata for Matching Patterns, in *Handbook of Formal Languages*, Volume 2, Linear Modeling: Background and Application, G. Rozenberg and A. Salomaa eds., Chapter 9, pp 399–462, Springer-Verlag, Berlin.

- CROCHEMORE, M., HANCART, C., LECROQ, T., 2001, *Algorithmique du texte*, Vuibert.

- CROCHEMORE, M., RYTTER, W., 1994, *Text Algorithms*, Oxford University Press.

- HANCART, C., 1992, Une analyse en moyenne de l'algorithme de Morris et Pratt et de ses raffinements, in *Théorie des Automates et Applications, Actes des 2e Journées Franco-Belges*, D. Krob ed., Rouen, France, pp 99–110, PUR 176, Rouen, France.

- HANCART, C., 1993, On Simon's string searching algorithm, *Information Processing Letters* **47**(2):95–99.

- HANCART, C., 1993, *Analyse exacte et en moyenne d'algorithmes de recherche d'un motif dans un texte*, Thèse de doctorat de l'Université de Paris 7, France.

- SIMON, I., 1993, String matching algorithms and automata, in *Proceedings of the 1st American Workshop on String Processing*, R.A. Baeza-Yates and N. Ziviani eds., pp 151–157, Universidade Federal de Minas Gerais, Brazil.

- SIMON, I., 1994, String matching algorithms and automata, in *Results and Trends in Theoretical Computer Science*, Graz, Austria, J. Karhumäki, H. Maurer and G. Rozenberg eds., pp 386–395, Lecture Notes in Computer Science 814, Springer-Verlag, Berlin.

CHAPTER 9

COLUSSI ALGORITHM

1 Main features

- refinement of the Knuth-Morris-Pratt algorithm;

- partitions the set of pattern positions into two disjoint subsets; the positions in the first set are scanned from left to right and when no mismatch occurs the positions of the second subset are scanned from right to left;

- preprocessing phase in $O(m)$ time and space complexity;

- searching phase in $O(n)$ time complexity;

- performs $\frac{3}{2}n$ text character comparisons in the worst case.

2 Description

The design of the Colussi algorithm follows a tight analysis of the Knuth-Morris-Pratt algorithm (see chapter 7).

The set of pattern positions is divided into two disjoint subsets. Then each attempt consists in two phases:

- in the first phase the comparisons are performed from left to right with text characters aligned with pattern position for which the value of the *kmpNext* function (see chapter 7) is strictly greater than -1. These positions are called **noholes**;

- the second phase consists in comparing the remaining positions (called **holes**) from right to left.

This strategy presents two advantages:

- when a mismatch occurs during the first phase, after the appropriate shift it is not necessary to compare the text characters aligned with noholes compared during the previous attempt;

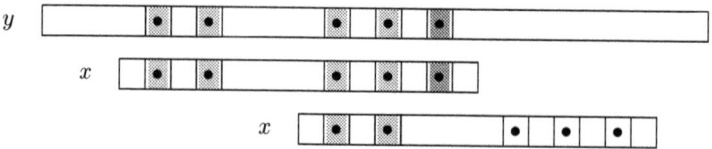

Figure 9.1. Mismatch with a nohole. Noholes are black circles and are compared from left to right. In this situation, after the shift, it is not necessary to compare the first two noholes again.

- when a mismatch occurs during the second phase it means that a suffix of the pattern matches a factor of the text, after the corresponding shift a prefix of the pattern will still match a factor of the text, then it is not necessary to compare this factor again.

For $0 \leq i \leq m-1$:

$$kmin[i] = \begin{cases} d > 0 & \text{if and only if } x[0..i-1-d] = x[d..i-1] \text{ and} \\ & x[i-d] \neq x[i], \\ 0 & \text{otherwise}. \end{cases}$$

When $kmin[i] \neq 0$ a periodicity ends at position i in x.

For $0 < i < m$ if $kmin[i-1] \neq 0$ then i is a nohole otherwise i is a hole.

Let $nd+1$ be the number of noholes in x. The table h contains first the $nd+1$ noholes in increasing order and then the $m-nd-1$ holes in decreasing order:

- for $0 \leq i \leq nd$, $h[i]$ is a nohole and $h[i] < h[i+1]$ for $0 \leq i < nd$;
- for $nd < i < m$, $h[i]$ is a hole and $h[i] > h[i+1]$ for $nd < i < m-1$.

If i is a hole then $rmin[i]$ is the smallest period of x greater than i.

The value of $first[u]$ is the smallest integer v such that $u \leq h[v]$.

Then assume that x is aligned with $y[j..j+m-1]$. If $x[h[k]] = y[j+h[k]]$ for $0 \leq k < r < nd$ and $x[h[r]] \neq y[j+h[r]]$. Let $j' = j + kmin[h[r]]$. Then there is no occurrence of x beginning in $y[j..j']$ and x can be shifted by $kmin[h[r]]$ positions to the right. Moreover $x[h[k]] = y[j' + h[k]]$ for $0 \leq k < first[h[r] - kmin[h[r]]]$ meaning that the comparisons can be resume with $x[h[first[h[r] - kmin[h[r]]]]]$ and $y[j' + h[first[h[r] - kmin[h[r]]]]]$ (see figure 9.1).

If $x[h[k]] = y[j + h[k]]$ for $0 \leq k < r$ and $x[h[r]] \neq y[j + h[r]]$ with $nd \leq r < m$. Let $j' = j + rmin[h[r]]$. Then there is no occurrence of

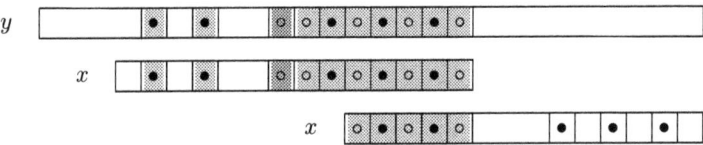

Figure 9.2. Mismatch with a hole. Noholes are black circles and are compared from left to right while holes are white circles and are compared from right to left. In this situation, after the shift, it is not necessary to compare the matched prefix of the pattern again.

x beginning in $y[j..j']$ and x can be shifted by $kmin[h[r]]$ positions to the right. Moreover $x[0..m-1-rmin[h[r]]] = y[j'..j+m-1]$ meaning that the comparisons can be resume with $x[h[first[m-1-rmin[h[r]]]]]$ and $y[j' + h[first[m-1-rmin[h[r]]]]]$ (see figure 9.2).

To compute the values of $kmin$, a table $hmax$ is used and defined as follows: $hmax[k]$ is such that $x[k..hmax[k]-1] = x[0..hmax[k]-k-1]$ and $x[hmax[k]] \neq x[hmax[k]-k]$.

The value of $ndh0[i]$ is the number of noholes strictly smaller than i.

We can now define two functions $shift$ and $next$ as follows:

- $shift[i] = kmin[h[i]]$ and $next[i] = ndh0[h[i] - kmin[h[i]]]$ for $i < nd$;

- $shift[i] = rmin[h[i]]$ and $next[i] = ndh0[m - rmin[h[i]]]$ for $nd \leq i < m$;

- $shift[m] = rmin[0]$ and $next[m] = ndh0[m - rmin[h[m-1]]]$.

Thus, during an attempt where the window is positioned on the text factor $y[j..j+m-1]$, when a mismatch occurs between $x[h[r]]$ and $y[j+h[r]]$ the window must be shifted by $shift[r]$ and the comparisons can be resume with pattern position $h[next[r]]$.

The preprocessing phase can be done in $O(m)$ space and time. The searching phase can then be done in $O(n)$ time complexity and furthermore at most $\frac{3}{2}n$ text character comparisons are performed during the searching phase.

3 The C code

```
int preColussi(String x, int m, int h[],
               int next[], int shift[]) {
  int i, k, nd, q, r, s;
  int hmax[XSIZE], kmin[XSIZE], nhd0[XSIZE], rmin[XSIZE];
```

```
/* Computation of hmax */
i = k = 1;
do {
   while (x[i] == x[i - k])
      i++;
   hmax[k] = i;
   q = k + 1;
   while (hmax[q - k] + k < i) {
      hmax[q] = hmax[q - k] + k;
      q++;
   }
   k = q;
   if (k == i + 1)
      i = k;
} while (k <= m);

/* Computation of kmin */
memset(kmin, 0, m*sizeof(int));
for (i = m; i >= 1; --i)
   if (hmax[i] < m)
      kmin[hmax[i]] = i;

/* Computation of rmin */
for (i = m - 1; i >= 0; --i) {
   if (hmax[i + 1] == m)
      r = i + 1;
   if (kmin[i] == 0)
      rmin[i] = r;
   else
      rmin[i] = 0;
}

/* Computation of h */
s = -1;
r = m;
for (i = 0; i < m; ++i)
   if (kmin[i] == 0)
      h[--r] = i;
   else
      h[++s] = i;
nd = s;
```

9. COLUSSI ALGORITHM

```
   /* Computation of shift */
   for (i = 0; i <= nd; ++i)
      shift[i] = kmin[h[i]];
   for (i = nd + 1; i < m; ++i)
      shift[i] = rmin[h[i]];
   shift[m] = rmin[0];

   /* Computation of nhd0 */
   s = 0;
   for (i = 0; i < m; ++i) {
      nhd0[i] = s;
      if (kmin[i] > 0)
         ++s;
   }

   /* Computation of next */
   for (i = 0; i <= nd; ++i)
      next[i] = nhd0[h[i] - kmin[h[i]]];
   for (i = nd + 1; i < m; ++i)
      next[i] = nhd0[m - rmin[h[i]]];
   next[m] = nhd0[m - rmin[h[m - 1]]];

   return(nd);
}

void COLUSSI(String x, int m, String y, int n) {
   int i, j, last, nd,
       h[XSIZE], next[XSIZE], shift[XSIZE];

   /* Processing */
   nd = preColussi(x, m, h, next, shift);

   /* Searching */
   i = j = 0;
   last = -1;
   while (j <= n - m) {
      while (i < m && last < j + h[i] &&
                      x[h[i]] == y[j + h[i]])
         i++;
      if (i >= m || last >= j + h[i]) {
         OUTPUT(j);
```

```
            i = m;
        }
        if (i > nd)
            last = j + m - 1;
        j += shift[i];
        i = next[i];
    }
}
```

4 The example

i	0	1	2	3	4	5	6	7	8
$x[i]$	G	C	A	G	A	G	A	G	
$kmpNext[i]$	-1	0	0	-1	1	-1	1	-1	1
$kmin[i]$	0	1	2	0	3	0	5	0	
$h[i]$	1	2	4	6	7	5	3	0	
$next[i]$	0	0	0	0	0	0	0	0	0
$shift[i]$	1	2	3	5	8	7	7	7	7
$hmax[i]$	0	1	2	4	4	6	6	8	8
$rmin[i]$	7	0	0	7	0	7	0	8	
$ndh0[i]$	0	0	1	2	2	3	3	4	

$nd = 3$

Searching phase

First attempt:

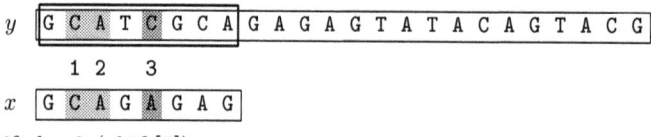

Shift by 3 ($shift[2]$)

Second attempt:

```
y  G C A T C G C A G A G A G T A T A C A G T A C G
         1 2
x        G C A G A G A G
```

Shift by 2 ($shift[1]$)

Third attempt:

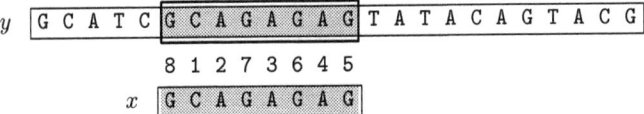

Shift by 7 (*shift*[8])

Fourth attempt:

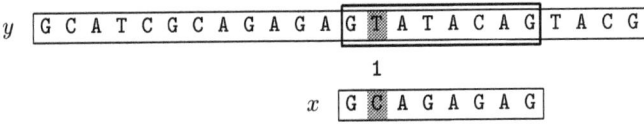

Shift by 1 (*shift*[0])

Fifth attempt:

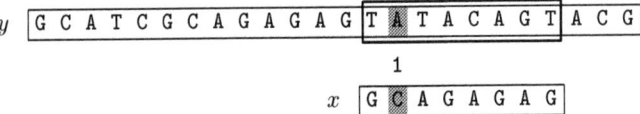

Shift by 1 (*shift*[0])

Sixth attempt:

```
y  G C A T C G C A G A G A G T A T A C A G T A C G
                                1
   x                       G C A G A G A G
```

Shift by 1 (*shift*[0])

Seventh attempt:

```
y  G C A T C G C A G A G A G T A T A C A G T A C G
                                  1
   x                         G C A G A G A G
```

Shift by 1 (*shift*[0])

Eighth attempt:

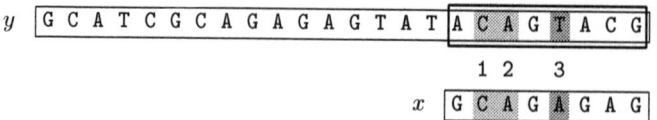

Shift by 3 (*shift*[2])

The Colussi algorithm performs 20 text character comparisons on the example.

5 References

- BRESLAUER, D., 1992, *Efficient String Algorithmics*, PhD Thesis, Report CU–024–92, Computer Science Department, Columbia University, New York, NY.

- COLUSSI, L., 1991, Correctness and efficiency of the pattern matching algorithms, *Information and Computation* **95**(2):225–251.

- COLUSSI, L., GALIL, Z., GIANCARLO, R., 1990, On the exact complexity of string matching, in *Proceedings of the 31st IEEE Annual Symposium on Foundations of Computer Science*, Saint Louis, MO, pp 135–144, IEEE Computer Society Press.

- GALIL, Z., GIANCARLO, R., 1992, On the exact complexity of string matching: upper bounds, *SIAM Journal on Computing*, **21** (3):407–437.

CHAPTER 10

GALIL-GIANCARLO ALGORITHM

1 Main features

- refinement of Colussi algorithm;
- preprocessing phase in $O(m)$ time and space complexity;
- searching phase in $O(n)$ time complexity;
- performs $\frac{4}{3}n$ text character comparisons in the worst case.

2 Description

The Galil-Giancarlo algorithm is a variant of the Colussi algorithm (see chapter 9). The change intervenes in the searching phase. The method applies when x is not a power of a single character. Thus $x \neq c^m$ with $c \in \Sigma$. Let ℓ be the last index in the pattern such that for $0 \leq i \leq \ell$, $x[0] = x[i]$ and $x[0] \neq x[\ell+1]$: $x = a^{\ell+1}bu$ for $a, b \in \Sigma$, $u \in \Sigma^*$ and $a \neq b$. Assume that during the previous attempt all the noholes have been matched and a suffix of the pattern has been matched meaning that after the corresponding shift a prefix of the pattern will still match a part of the text. Thus the window is positioned on the text factor $y[j \mathinner{.\,.} j + m - 1]$ and the portion $y[j \mathinner{.\,.} \mathit{last}]$ matches $x[0 \mathinner{.\,.} \mathit{last} - j]$. Then during the next attempt the algorithm will scanned the text character beginning with $y[\mathit{last}+1]$ until either the end of the text is reached or a character $x[0] \neq y[j+k]$ is found. In this latter case two subcases can arise:

- $x[\ell+1] \neq y[j+k]$ or too less $x[0]$ have been found ($k \leq \ell$) then the window is shifted and positioned on the text factor $y[k+1 \mathinner{.\,.} k+m]$, the scanning of the text is resumed (as in the Colussi algorithm) with the first nohole and the memorized prefix of the pattern is the empty word.

- $x[\ell+1] = y[j+k]$ and enough of $x[0]$ has been found ($k > \ell$) then the window is shifted and positioned on the text factor $y[k - \ell - 1 \mathinner{.\,.} k - \ell + m - 2]$, the scanning of the text is resumed (as in the Colussi algorithm) with the second nohole ($x[\ell + 1]$ is the first one) and the memorized prefix of the pattern is $x[0 \mathinner{.\,.} \ell + 1]$.

The preprocessing phase is exactly the same as in the Colussi algorithm (chapter 9) and can be done in $O(m)$ space and time. The searching phase can then be done in $O(n)$ time complexity and furthermore at most $\frac{4}{3}n$ text character comparisons are performed during the searching phase.

3 The C code

The function preColussi is given chapter 9.

```
void GG(String x, int m, String y, int n) {
   int i, j, k, ell, last, nd;
   int h[XSIZE], next[XSIZE], shift[XSIZE];
   char heavy;

   for (ell = 0; x[ell] == x[ell + 1]; ell++);
   if (ell == m - 1)
      /* Searching for a power of a single character */
      for (j = ell = 0; j < n; ++j)
         if (x[0] == y[j]) {
            ++ell;
            if (ell >= m)
               OUTPUT(j - m + 1);
         }
         else
            ell = 0;
   else {
      /* Preprocessing */
      nd = preCOLUSSI(x, m, h, next, shift);
      /* Searching */
      i = j = heavy = 0;
      last = -1;
      while (j <= n - m) {
         if (heavy && i == 0) {
            k = last - j + 1;
            while (x[0] == y[j + k])
               k++;
            if (k <= ell || x[ell + 1] != y[j + k]) {
               i = 0;
               j += (k + 1);
               last = j - 1;
```

```
            }
            else {
               i = 1;
               last = j + k;
               j = last - (ell + 1);
            }
            heavy = 0;
         }
         else {
            while (i < m && last < j + h[i] &&
                              x[h[i]] == y[j + h[i]])
               ++i;
            if (i >= m || last >= j + h[i]) {
               OUTPUT(j);
               i = m;
            }
            if (i > nd)
               last = j + m - 1;
            j += shift[i];
            i = next[i];
         }
         heavy = (j > last ? 0 : 1);
      }
   }
}
```

4   The example

i	0	1	2	3	4	5	6	7	8
$x[i]$	G	C	A	G	A	G	A	G	
$kmpNext[i]$	-1	0	0	-1	1	-1	1	-1	1
$kmin[i]$	0	1	2	0	3	0	5	0	
$h[i]$	1	2	4	6	7	5	3	0	
$next[i]$	0	0	0	0	0	0	0	0	0
$shift[i]$	1	2	3	5	8	7	7	7	7
$hmax[i]$	0	1	2	4	4	6	6	8	8
$rmin[i]$	7	0	0	7	0	7	0	8	
$ndh0[i]$	0	0	1	2	2	3	3	4	

$nd = 3$ and $\ell = 0$

Searching phase

First attempt:

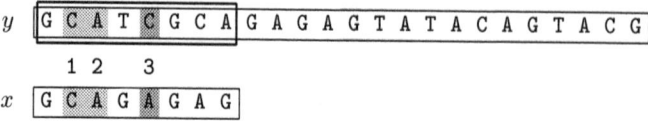

Shift by 3 (*shift*[2])

Second attempt:

```
y  G C A T C G C A G A G A G T A T A C A G T A C G
         1 2
   x   G C A G A G A G
```

Shift by 2 (*shift*[1])

Third attempt:

```
y  G C A T C G C A G A G A G T A T A C A G T A C G
             8 1 2 7 3 6 4 5
         x   G C A G A G A G
```

Shift by 7 (*shift*[8])

Fourth attempt:

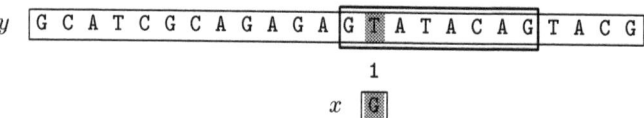

Shift by 2

Fifth attempt:

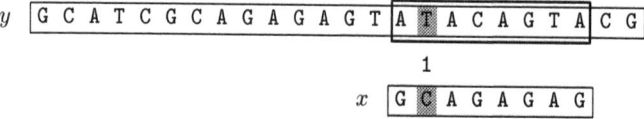

Shift by 1 (*shift*[0])

Sixth attempt:

```
y  G C A T C G C A G A G A G T A T A C A G T A C G
                                  1
x                                G C A G A G A G
```

Shift by 1 (*shift*[0])

Seventh attempt:

```
y  G C A T C G C A G A G A G T A T A C A G T A C G
                                  1 2   3
x                                G C A G A G A G
```

Shift by 3 (*shift*[2])

The Galil-Giancarlo algorithm performs 19 text character comparisons on the example.

5 References

- BRESLAUER, D., 1992, *Efficient String Algorithmics*, PhD Thesis, Report CU–024–92, Computer Science Department, Columbia University, New York, NY.

- GALIL, Z., GIANCARLO, R., 1992, On the exact complexity of string matching: upper bounds, *SIAM Journal on Computing*, **21** (3):407–437.

CHAPTER 11

APOSTOLICO-CROCHEMORE ALGORITHM

1 Main features

- preprocessing phase in $O(m)$ time and space complexity;
- searching phase in $O(n)$ time complexity;
- performs $\frac{3}{2}n$ text character comparisons in the worst case.

2 Description

The Apostolico-Crochemore uses the *kmpNext* shift table (see chapter 7) to compute the shifts. Let $\ell = 0$ if x is a power of a single character ($x = c^m$ with $c \in \Sigma$) and ℓ be equal to the position of the first character of x different from $x[0]$ otherwise ($x = a^\ell bu$ for $a, b \in \Sigma$, $u \in \Sigma^*$ and $a \neq b$). During each attempt the comparisons are made with pattern positions in the following order: $\ell, \ell + 1, \ldots, m - 2, m - 1, 0, 1, \ldots, \ell - 1$. During the searching phase we consider triple of the form (i, j, k) where:

- the window is positioned on the text factor $y[j \mathinner{.\,.} j + m - 1]$;
- $0 \leq k \leq \ell$ and $x[0 \mathinner{.\,.} k - 1] = y[j \mathinner{.\,.} j + k - 1]$;
- $\ell \leq i \leq m$ and $x[\ell \mathinner{.\,.} i - 1] = y[j + \ell \mathinner{.\,.} i + j - 1]$.

(see figure 11.1).

The initial triple is $(\ell, 0, 0)$.

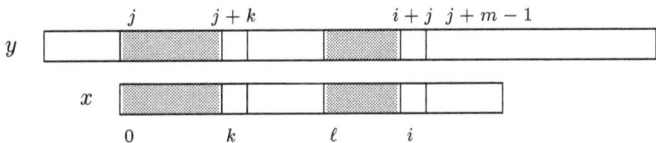

Figure 11.1. At each attempt of the Apostolico-Crochemore algorithm we consider a triple (i, j, k).

We now explain how to compute the next triple after (i,j,k) has been computed. Three cases arise depending on the value of i:

- $i = \ell$

  If $x[i] = y[i+j]$ then the next triple is $(i+1, j, k)$.

  If $x[i] \neq y[i+j]$ then the next triple is $(\ell, j+1, \max\{0, k-1\})$.

- $\ell < i < m$

  If $x[i] = y[i+j]$ then the next triple is $(i+1, j, k)$.

  If $x[i] \neq y[i+j]$ then two cases arise depending on the value of kmpNext[i]:

  - kmpNext[i] $\leq \ell$: then the next triple is

    $$(\ell, i+j - \text{kmpNext}[i], \max\{0, \text{kmpNext}[i]\}),$$

  - kmpNext[i] $> \ell$: then the next triple is

    $$(\text{kmpNext}[i], i+j - \text{kmpNext}[i], \ell).$$

- $i = m$

  If $k < \ell$ and $x[k] = y[j+k]$ then the next triple is $(i, j, k+1)$.

  Otherwise either $k < \ell$ and $x[k] \neq y[j+k]$, or $k = \ell$. If $k = \ell$ an occurrence of x is reported. In both cases the next triple is computed in the same manner as in the case where $\ell < i < m$.

The preprocessing phase consists in computing the table kmpNext and the integer ℓ. It can be done in $O(m)$ space and time. The searching phase is in $O(n)$ time complexity and furthermore the Apostolico-Crochemore algorithm performs at most $\frac{3}{2}n$ text character comparisons in the worst case.

3 The C code

The function preKmp is given chapter 7.

```
void AXAMAC(String x, int m, String y, int n) {
   int i, j, k, ell, kmpNext[XSIZE];

   /* Preprocessing */
   preKmp(x, m, kmpNext);
   for (ell = 1; x[ell - 1] == x[ell]; ell++);
```

```
      if (ell == m)
         ell = 0;

      /* Searching */
      i = ell;
      j = k = 0;
      while (j <= n - m) {
         while (i < m && x[i] == y[i + j])
            ++i;
         if (i >= m) {
            while (k < ell && x[k] == y[j + k])
               ++k;
            if (k >= ell)
               OUTPUT(j);
         }
         j += (i - kmpNext[i]);
         if (i == ell)
            k = MAX(0, k - 1);
         else
            if (kmpNext[i] <= ell) {
               k = MAX(0, kmpNext[i]);
               i = ell;
            }
            else {
               k = ell;
               i = kmpNext[i];
            }
      }
   }
```

4 The example

i	0	1	2	3	4	5	6	7	8
$x[i]$	G	C	A	G	A	G	A	G	
$kmpNext[i]$	−1	0	0	−1	1	−1	1	−1	1

$\ell = 1$

Searching phase

First attempt:

```
y  G C A T C G C A G A G A G T A T A C A G T A C G
       1 2 3
x  G C A G A G A G
```

Shift by 4 ($i - kmpNext[i] = 3 - -1$)

Second attempt:

```
y  G C A T C G C A G A G A G T A T A C A G T A C G
             1
       x  G C A G A G A G
```

Shift by 1 ($i - kmpNext[i] = 1 - 0$)

Third attempt:

```
y  G C A T C G C A G A G A G T A T A C A G T A C G
             8 1 2 3 4 5 6 7
          x  G C A G A G A G
```

Shift by 7 ($i - kmpNext[i] = 8 - 1$)

Fourth attempt:

```
y  G C A T C G C A G A G A G T A T A C A G T A C G
                             1
                   x  G C A G A G A G
```

Shift by 1 ($i - kmpNext[i] = 1 - 0$)

Fifth attempt:

```
y  G C A T C G C A G A G A G T A T A C A G T A C G
                               1
                     x  G C A G A G A G
```

Shift by 1 ($i - kmpNext[i] = 0 - -1$)

Sixth attempt:

| y | G C A T C G C A G A G A G T A T A C A G T A C G |

1

| x | G C A G A G A G |

Shift by 1 ($i - kmpNext[i] = 0 - -1$)

Seventh attempt:

| y | G C A T C G C A G A G A G T A T A C A G T A C G |

1

| x | G C A G A G A G |

Shift by 1 ($i - kmpNext[i] = 0 - -1$)

Eighth attempt:

| y | G C A T C G C A G A G A G T A T A C A G T A C G |

1 2 3 4

| x | G C A G A G A G |

Shift by 3 ($i - kmpNext[i] = 4 - 1$)

On the example the Apostolico-Crochemore algorithm performs 20 text character comparisons on the example.

5 References

- APOSTOLICO, A., CROCHEMORE, M., 1991, Optimal canonization of all substrings of a string, *Information and Computation* **95**(1):76–95.

- HANCART, C., 1993, *Analyse exacte et en moyenne d'algorithmes de recherche d'un motif dans un texte*, Thèse de doctorat de l'Université de Paris 7, France.

- SMYTH, W. F., 2003, *Computing Patterns in Strings*, Pearson Addison Wesley.

CHAPTER 12

NOT SO NAIVE ALGORITHM

1 Main features

- preprocessing phase in constant time complexity;
- constant extra space complexity;
- searching phase in $O(m \times n)$ time complexity;
- (slightly) sub-linear in the average case.

2 Description

During the searching phase of the Not So Naive algorithm the character comparisons are made with the pattern positions in the following order $1, 2, \ldots, m-2, m-1, 0$.

For each attempt where the window is positioned on the text factor $y[j \mathinner{.\,.} j + m - 1]$: if $x[0] = x[1]$ and $x[1] \neq y[j+1]$ or if $x[0] \neq x[1]$ and $x[1] = y[j+1]$ the pattern is shifted by 2 positions at the end of the attempt and by 1 otherwise.

Thus the preprocessing phase can be done in constant time and space. The searching phase of the Not So Naive algorithm has a quadratic worst case but it is slightly sub-linear in the average case.

3 The C code

```
void NSN(String x, int m, String y, int n) {
   int j, k, ell;

   /* Preprocessing */
   if (x[0] == x[1]) {
      k = 2; ell = 1;
   }
   else {
      k = 1; ell = 2;
   }
```

```
/* Searching */
j = 0;
while (j <= n - m)
    if (x[1] != y[j + 1])
        j += k;
    else {
        if (memcmp(x + 2, y + j + 2, m - 2) == 0 &&
            x[0] == y[j])
            OUTPUT(j);
        j += ell;
    }
}
```

4 The example

$k = 1$ and $\ell = 2$

Searching phase

First attempt:

y | G **CA T** C G C A G A G A G T A T A C A G T A C G
        1 2 3
x | G **CA G** A G A G

Shift by 2

Second attempt:

y | G C **A T** C G C A G A G A G T A T A C A G T A C G
          1
x | G **C** A G A G A G

Shift by 1

Third attempt:

y | G C A T **C G** C A G A G A G T A T A C A G T A C G
            1 2
x | G **C A** G A G A G

Shift by 2

12. NOT SO NAIVE ALGORITHM

Fourth attempt:

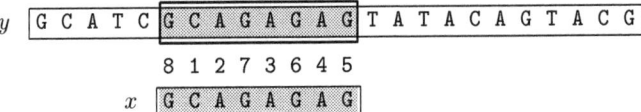

Shift by 2

Fifth attempt:

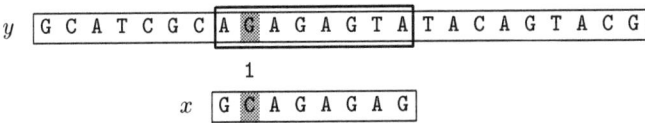

Shift by 1

Sixth attempt:

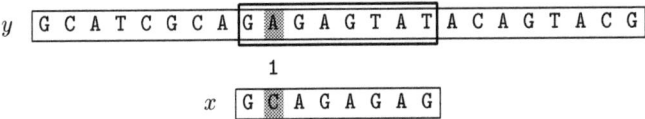

Shift by 1

Seventh attempt:

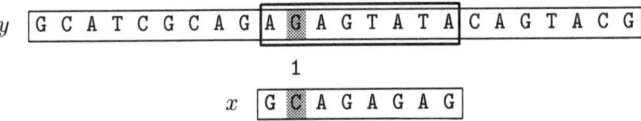

Shift by 1

Eighth attempt:

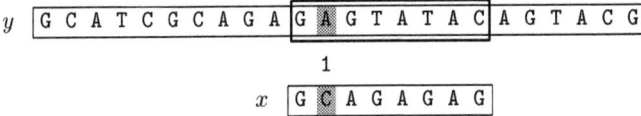

Shift by 1

Ninth attempt:

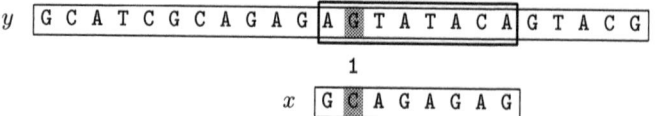

Shift by 1

Tenth attempt:

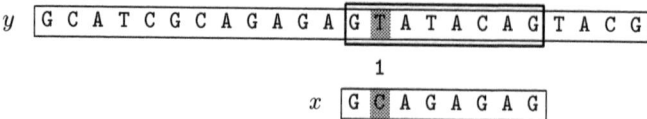

Shift by 1

Eleventh attempt:

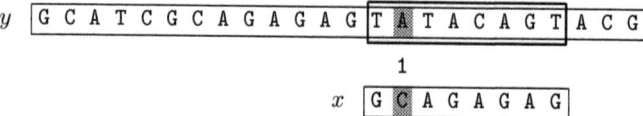

Shift by 1

Twelfth attempt:

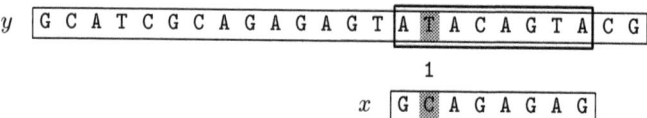

Shift by 1

Thirteenth attempt:

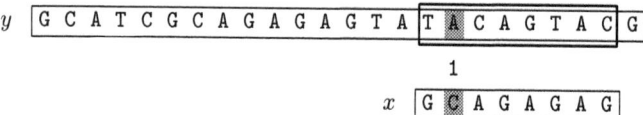

Shift by 1

Fourteenth attempt:

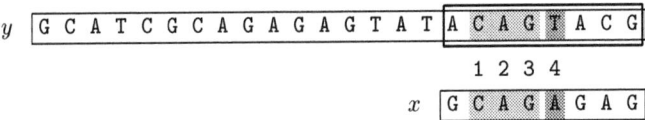

Shift by 2

The Not So Naive algorithm performs 27 text character comparisons on the example.

5 References

- CARDON, A., CHARRAS, C., 1996, *Introduction à l'algorithmique et à la programmation*, Chapter 9, pp 254–279, Ellipses.

- HANCART, C., 1992, Une analyse en moyenne de l'algorithme de Morris et Pratt et de ses raffinements, in *Théorie des Automates et Applications, Actes des 2e Journées Franco-Belges*, D. Krob ed., Rouen, France, pp 99–110, PUR 176, Rouen, France.

- HANCART, C., 1993, *Analyse exacte et en moyenne d'algorithmes de recherche d'un motif dans un texte*, Thèse de doctorat de l'Université de Paris 7, France.

CHAPTER 13

FORWARD DAWG MATCHING ALGORITHM

1 Main Features

- uses the suffix automaton of x;
- $O(n)$ worst case time complexity;
- performs exactly n text character inspections.

2 Description

The Forward Dawg Matching algorithm computes the longest factor of the pattern ending at each position in the text. This is make possible by the use of the smallest suffix automaton (also called DAWG for Directed Acyclic Word Graph) of the pattern. The smallest suffix automaton of a word w is a Deterministic Finite Automaton $\mathcal{S}(w) = (Q, q_0, T, E)$. The language accepted by $\mathcal{S}(w)$ is $\mathcal{L}(\mathcal{S}(w)) = \{u \in \Sigma^* \mid \exists v \in \Sigma^* \text{ such that } w = vu\}$. The preprocessing phase of the Forward Dawg Matching algorithm consists in computing the smallest suffix automaton for the pattern x. It is linear in time and space in the length of the pattern.

During the searching phase the Forward Dawg Matching algorithm parses the characters of the text from left to right with the automaton $\mathcal{S}(x)$ starting with state q_0. For each state $q \in \mathcal{S}(x)$ the length of the longest path from q_0 to p is denoted by $length(q)$. This structure extensively uses the notion of suffix links. For each state p the suffix link of p is denoted by $S[p]$. For a state p, let $Path(p) = (p_0, p_1, \ldots, p_\ell)$ be the suffix path of p such that $p_0 = p$, for $1 \leq i \leq \ell$, $p_i = S[p_{i-1}]$ and $p_\ell = q_0$. For each text character $y[j]$ sequentially, let p be the current state, then the Forward Dawg Matching algorithm takes a transition defined for $y[j]$ for the first state of $Path(p)$ for which such a transition is defined. The current state p is updated with the target state of this transition or with the initial state q_0 if no transition exists labeled with $y[j]$ from a state of $Path(p)$. An occurrence of x is found when $length(p) = m$.

The Forward Dawg Matching algorithm performs exactly n text character inspections.

3   The C code

The function buildSuffixAutomaton is given chapter 29. All the other functions to build and manipulate the suffix automaton can be found section 5.

```
int FDM(String x, int m, String y, int n) {
   int j, init, ell, state;
   Graph aut;

   /* Preprocessing */
   aut = newSuffixAutomaton(2*(m + 2), 2*(m + 2)*ASIZE);
   buildSuffixAutomaton(x, m, aut);
   init = getInitial(aut);

   /* Searching */
   ell = 0;
   state = init;
   for (j = 0; j < n; ++j) {
      if (getTarget(aut, state, y[j]) != UNDEFINED) {
         ++ell;
         state = getTarget(aut, state, y[j]);
      }
      else {
         while (state != init &&
                getTarget(aut, state, y[j]) == UNDEFINED)
            state = getSuffixLink(aut, state);
         if (getTarget(aut, state, y[j]) != UNDEFINED) {
            ell = getLength(aut, state) + 1;
            state = getTarget(aut, state, y[j]);
         }
         else {
            ell = 0;
            state = init;
         }
      }
      if (ell == m)
         OUTPUT(j - m + 1);
   }
}
```

4 The example

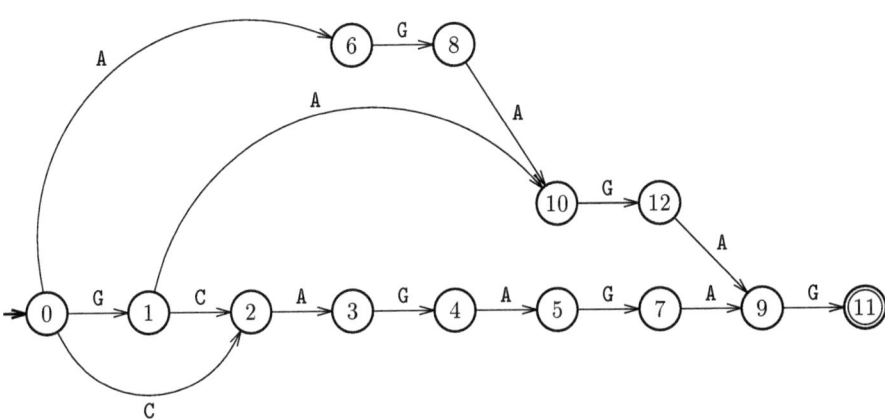

state	0	1	2	3	4	5	6	7	8	9	10	11	12
suffix link	0	0	0	6	8	10	0	12	1	10	6	12	8
length	0	1	2	3	4	5	1	6	2	7	3	8	4

Searching phase

The initial state is 0.

y G C A T C G C A G A G A G T A T A C A G T A C G
   1 2 3 0

y G C A T C G C A G A G A G T A T A C A G T A C G
         2 0

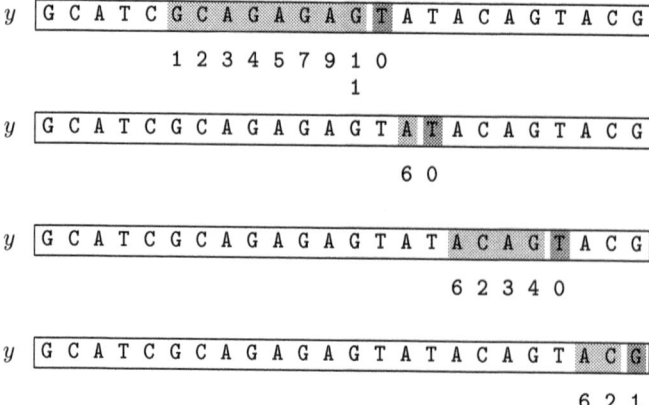

The Forward Dawg Matching algorithm performs exactly 24 text character inspections on the example.

5 References

- CROCHEMORE, M., RYTTER, W., 1994, *Text Algorithms*, Oxford University Press.

CHAPTER 14

BOYER-MOORE ALGORITHM

1 Main Features

- performs the comparisons from right to left;
- preprocessing phase in $O(m + \sigma)$ time and space complexity;
- searching phase in $O(m \times n)$ time complexity;
- $3n$ text character comparisons in the worst case when searching for a non periodic pattern;
- $O(n/m)$ best performance.

2 Description

The Boyer-Moore algorithm is considered as the most efficient string matching algorithm in usual applications. A simplified version of it or the entire algorithm is often implemented in text editors for the "search" and "substitute" commands.

The algorithm scans the characters of the pattern from right to left beginning with the rightmost one. In case of a mismatch (or a complete match of the whole pattern) it uses two precomputed functions to shift the window to the right. These two shift functions are called the **good-suffix shift** (also called matching shift) and the **bad-character shift** (also called indexoccurrence shiftoccurrence shift).

Assume that a mismatch occurs between the character $x[i] = a$ of the pattern and the character $y[i+j] = b$ of the text during an attempt at the left position j. Then, $x[i+1 .. m-1] = y[i+j+1 .. j+m-1] = u$ and $x[i] \neq y[i+j]$. The good-suffix shift consists in aligning the segment $y[i+j+1 .. j+m-1] = x[i+1 .. m-1]$ with its rightmost occurrence in x that is preceded by a character different from $x[i]$ (see figure 14.1). If there exists no such segment, the shift consists in aligning the longest suffix v of $y[i+j+1 .. j+m-1]$ with a matching prefix of x (see figure 14.2).

The bad-character shift consists in aligning the text character $y[i+j]$ with its rightmost occurrence in $x[0 .. m-2]$ (see figure 14.3). If $y[i+j]$

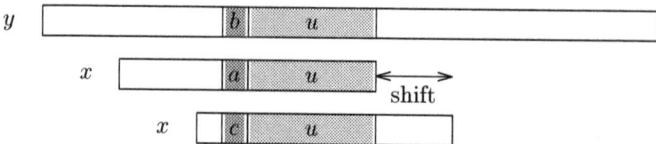

Figure 14.1. The good-suffix shift, u re-occurs preceded by a character c different from a.

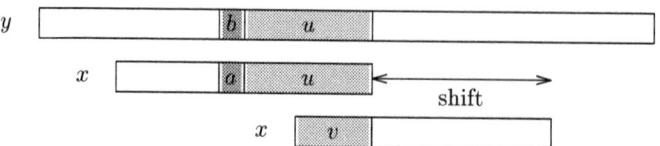

Figure 14.2. The good-suffix shift, only a suffix of u re-occurs in x.

does not occur in the pattern x, no occurrence of x in y can include $y[i+j]$, and the left end of the window is aligned with the character immediately after $y[i+j]$, namely $y[i+j+1]$ (see figure 14.4).

Note that the bad-character shift can be negative, thus for shifting the window, the Boyer-Moore algorithm applies the maximum between the good-suffix shift and bad-character shift. More formally the two shift functions are defined as follows.

The good-suffix shift function is stored in a table $bmGs$ of size $m+1$. Let us define two conditions:

$Cs(i, s)$: for each k such that $i < k < m, s \geq k$ or $x[k-s] = x[k]$,

and

$Co(i, s)$: if $s < i$ then $x[i-s] \neq x[i]$.

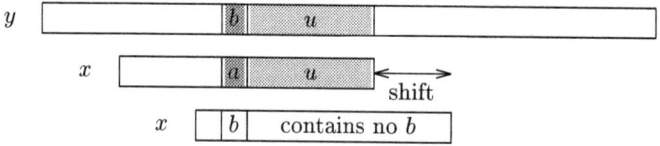

Figure 14.3. The bad-character shift, b occurs in x.

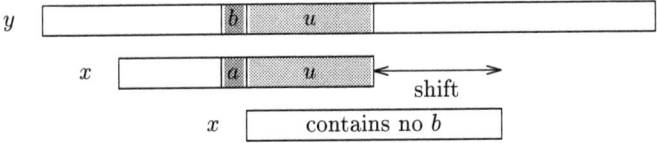

Figure 14.4. The bad-character shift, b does not occur in x.

Then, for $0 \leq i < m$:

$$bmGs[i+1] = \min\{s > 0 \mid Cs(i,s) \text{ and } Co(i,s) \text{ hold}\}$$

and we define $bmGs[0]$ as the length of the period of x. The computation of the table $bmGs$ use a table suff defined as follows:

for $1 \leq i < m$, $\mathit{suff}[i] = \max\{k \mid x[i-k+1..i] = x[m-k, m-1]\}$.

The bad-character shift function is stored in a table $bmBc$ of size σ. For $c \in \Sigma$:

$$bmBc[c] = \begin{cases} \min\{i \mid 1 \leq i < m-1 \text{ and } x[m-1-i] = c\} & if c occurs \\ & \text{in } x, \\ m & \text{otherwise}. \end{cases}$$

Tables $bmBc$ and $bmGs$ can be precomputed in time $O(m+\sigma)$ before the searching phase and require an extra-space in $O(m+\sigma)$. The searching phase time complexity is quadratic but at most $3n$ text character comparisons are performed when searching for a non periodic pattern. On large alphabets (relatively to the length of the pattern) the algorithm is extremely fast. When searching for a^m in $(a^{m-1}b)^{n/m}$ the algorithm makes only $O(n/m)$ comparisons, which is the absolute minimum for any string matching algorithm in the model where only the pattern is preprocessed.

3 The C code

```
void preBmBc(String x, int m, int bmBc[]) {
   int i;

   for (i = 0; i < ASIZE; ++i)
      bmBc[i] = m;
   for (i = 0; i < m - 1; ++i)
      bmBc[x[i]] = m - i - 1;
}
```

```
void suffixes(String x, int m, int *suff) {
   int f, g, i;

   suff[m - 1] = m;
   g = m - 1;
   for (i = m - 2; i >= 0; --i) {
      if (i > g && suff[i + m - 1 - f] < i - g)
         suff[i] = suff[i + m - 1 - f];
      else {
         if (i < g)
            g = i;
         f = i;
         while (g >= 0 && x[g] == x[g + m - 1 - f])
            --g;
         suff[i] = f - g;
      }
   }
}

void preBmGs(String x, int m, int bmGs[]) {
   int i, j, suff[XSIZE];

   suffixes(x, m, suff);

   for (i = 0; i < m; ++i)
      bmGs[i] = m;
   j = 0;
   for (i = m - 1; i >= -1; --i)
      if (i == -1 || suff[i] == i + 1)
         for (; j < m - 1 - i; ++j)
            if (bmGs[j] == m)
               bmGs[j] = m - 1 - i;
   for (i = 0; i <= m - 2; ++i)
      bmGs[m - 1 - suff[i]] = m - 1 - i;
}
```

```
void BM(String x, int m, String y, int n) {
   int i, j, bmGs[XSIZE], bmBc[ASIZE];

   /* Preprocessing */
   preBmGs(x, m, bmGs);
   preBmBc(x, m, bmBc);

   /* Searching */
   j = 0;
   while (j <= n - m) {
      for (i = m - 1; i >= 0 && x[i] == y[i + j]; --i);
      if (i < 0) {
         OUTPUT(j);
         j += bmGs[0];
      }
      else
         j += MAX(bmGs[i], bmBc[y[i + j]] - m + 1 + i);
   }
}
```

4 The example

c	A	C	G	T
$bmBc[c]$	1	6	2	8

i	0	1	2	3	4	5	6	7
$x[i]$	G	C	A	G	A	G	A	G
$\text{suff}[i]$	1	0	0	2	0	4	0	8
$bmGs[i]$	7	7	7	2	7	4	7	1

Searching phase

First attempt:

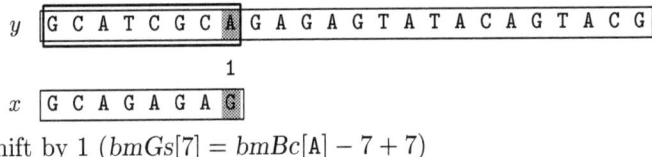

Shift by 1 ($bmGs[7] = bmBc[\text{A}] - 7 + 7$)

Second attempt:

```
y  G C A T C G C A G A G A G T A T A C A G T A C G
             3 2 1
x     G C A G A G A G
```
Shift by 4 ($bmGs[5] = bmBc[C] - 7 + 5$)

Third attempt:

```
y  G C A T C G C A G A G A G T A T A C A G T A C G
         8 7 6 5 4 3 2 1
x         G C A G A G A G
```
Shift by 7 ($bmGs[0]$)

Fourth attempt:

```
y  G C A T C G C A G A G A G T A T A C A G T A C G
                         3 2 1
x             G C A G A G A G
```
Shift by 4 ($bmGs[5] = bmBc[C] - 7 + 5$)

Fifth attempt:

```
y  G C A T C G C A G A G A G T A T A C A G T A C G
                                       2 1
x                 G C A G A G A G
```
Shift by 7 ($bmGs[6]$)

The Boyer-Moore algorithm performs 17 text character comparisons on the example.

5 References

- AHO, A.V., 1990, Algorithms for Finding Patterns in Strings, in *Handbook of Theoretical Computer Science, Volume A, Algorithms and complexity*, J. van Leeuwen ed., Chapter 5, pp 255–300, Elsevier, Amsterdam.

- AOE, J.-I., 1994, *Computer algorithms: string pattern matching strategies*, IEEE Computer Society Press.

- BAASE, S., VAN GELDER, A., 1999, *Computer Algorithms: Introduction to Design and Analysis*, 3rd Edition, Chapter 11, Addison-Wesley Publishing Company.

- BAEZA-YATES, R.A., NAVARRO G., RIBEIRO-NETO B., 1999, Indexing and Searching, in *Modern Information Retrieval*, Chapter 8, pp 191–228, Addison-Wesley.

- BEAUQUIER, D., BERSTEL, J., CHRÉTIENNE, P., 1992, *Éléments d'algorithmique*, Chapter 10, pp 337–377, Masson, Paris.

- BINSTOCK, A., REX, J., 1995, *Practical algorithms for programmers*, Addison Wesley.

- BOYER, R.S., MOORE, J.S., 1977, A fast string searching algorithm, *Communications of the ACM*. 20:762–772.

- COLE, R., 1994, Tight bounds on the complexity of the Boyer–Moore pattern matching algorithm, *SIAM Journal on Computing* 23(5):1075-1091.

- CORMEN, T.H., LEISERSON, C.E., RIVEST, R.L., 1990, *Introduction to Algorithms*, Chapter 34, pp 853–885, MIT Press.

- CROCHEMORE, M., 1997, Off-line serial exact string searching, in *Pattern Matching Algorithms*, A. Apostolico and Z. Galil eds., Chapter 1, pp 1–53, Oxford University Press.

- CROCHEMORE, M., HANCART, C., 1999, Pattern Matching in Strings, in *Algorithms and Theory of Computation Handbook*, M.J. Atallah ed., Chapter 11, pp 11-1–11-28, CRC Press Inc., Boca Raton, FL.

- CROCHEMORE, M., HANCART, C., LECROQ, T., 2001, *Algorithmique du texte*, Vuibert.

- CROCHEMORE, M., LECROQ, T., 1996, Pattern matching and text compression algorithms, in *CRC Computer Science and Engineering Handbook*, A.B. Tucker Jr ed., Chapter 8, pp 162–202, CRC Press Inc., Boca Raton, FL.

- CROCHEMORE, M., RYTTER, W., 1994, *Text Algorithms*, Oxford University Press.

- CROCHEMORE, M., RYTTER, W., 2002, *Jewels of Stringology*, World Scientific Press.

- GONNET, G.H., BAEZA-YATES, R.A., 1991, *Handbook of Algorithms and Data Structures in Pascal and C*, 2nd Edition, Chapter 7, pp. 251–288, Addison-Wesley Publishing Company.
- GOODRICH, M.T., TAMASSIA, R., 1998, *Data Structures and Algorithms in JAVA*, Chapter 11, pp 441–467, John Wiley & Sons.
- GUSFIELD, D., 1997, *Algorithms on strings, trees, and sequences: Computer Science and Computational Biology*, Cambridge University Press.
- HANCART, C., 1993, *Analyse exacte et en moyenne d'algorithmes de recherche d'un motif dans un texte*, Thèse de doctorat de l'Université de Paris 7, France.
- KNUTH, D.E., MORRIS, JR, J.H., PRATT, V.R., 1977, Fast pattern matching in strings, *SIAM Journal on Computing* **6**(1):323-350.
- LECROQ, T., 1992, *Recherches de mot*, Thèse de doctorat de l'Université d'Orléans, France.
- LECROQ, T., 1995, Experimental results on string matching algorithms, *Software – Practice & Experience* **25**(7):727-765.
- LECROQ, T., 2000, *Quelques aspects de l'algorithmique du texte*, Habilitation thesis, Université de Rouen, France.
- NAVARRO, G., RAFFINOT, M., 2002, *Flexible Pattern Matching in Strings Practical on-line search algorithms for texts and biological sequences* , Cambridge University Press.
- SEDGEWICK, R., 1988, *Algorithms*, Chapter 19, pp. 277–292, Addison-Wesley Publishing Company.
- SEDGEWICK, R., 1992, *Algorithms in C*, Chapter 19, Addison-Wesley Publishing Company.
- SMYTH, W. F., 2003, *Computing Patterns in Strings*, Pearson Addison Wesley.
- STEPHEN, G.A., 1994, *String Searching Algorithms*, World Scientific.
- WATSON, B.W., 1995, *Taxonomies and Toolkits of Regular Language Algorithms*, PhD Thesis, Eindhoven University of Technology, The Netherlands.
- WIRTH, N., 1986, *Algorithms & Data Structures*, Chapter 1, pp. 17–72, Prentice-Hall.

CHAPTER 15

GALIL ALGORITHM

1 Main Features

- improvement of the Boyer-Moore algorithm for periodic pattern;
- preprocessing phase in $O(m + \sigma)$ time and space complexity;
- searching phase in $O(n)$ time complexity;
- $14n$ text character comparisons in the worst case;
- $O(n/m)$ best performance.

2 Description

The Boyer-Moore algorithm (see chapter 14) has a quadratic time behavior for periodic pattern. In that case it is enough to slightly modify the Boyer-Moore algorithm to make it linear. Indeed, the index i on the pattern can go from $m-1$ to 0 at each attempt, unless an occurrence of the pattern has been found in the previous attempt. If it is the case, the index i will go from $m-1$ to $m - per(x)$. This technique is known as the "memorization prefix" technique.

3 The C code

The functions `preBmBc` and `preBmGs` are given chapter 14.

At each attempt, the variable `ell` memorizes the left bound of the variable i: either 0 if no occurrence of the pattern has been found in the previous attempt or $m - per(x)$ otherwise.

```
void GALIL(String x, int m, String y, int n) {
   int i, j, ell, bmGs[XSIZE], bmBc[ASIZE];

   /* Preprocessing */
   preBmGs(x, m, bmGs);
   preBmBc(x, m, bmBc);
```

```
/* Searching */
j = 0;
ell = 0;
while (j <= n - m) {
   for (i = m - 1; i >= ell && x[i] == y[i + j]; --i);
   if (i < ell) {
      OUTPUT(j);
      j += bmGs[0];
      ell = m - bmGs[0];
   }
   else {
      j += MAX(bmGs[i], bmBc[y[i + j]] - m + 1 + i);
      ell = 0;
   }
}
}
```

4 The example

c	A	C	G	T
$bmBc[c]$	1	6	2	8

i	0	1	2	3	4	5	6	7
$x[i]$	G	C	A	G	A	G	A	G
$suff[i]$	1	0	0	2	0	4	0	8
$bmGs[i]$	7	7	7	2	7	4	7	1

Searching phase

First attempt:

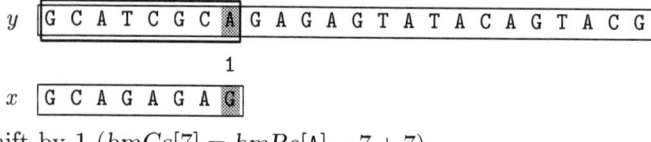

Shift by 1 ($bmGs[7] = bmBc[\text{A}] - 7 + 7$)

Second attempt:

y | G | C | A | T | C | G | C | A | G | A | G | A | G | T | A | T | A | C | A | G | T | A | C | G |

         3 2 1

x | G C A G A G A G |

Shift by 4 ($bmGs[5] = bmBc[\text{C}] - 7 + 5$)

Third attempt:

```
y  G C A T C G C A G A G A G T A T A C A G T A C G
             8 7 6 5 4 3 2 1
         x   G C A G A G A G
```

Shift by 7 ($bmGs[0]$)

Fourth attempt:

```
y  G C A T C G C A G A G A G T A T A C A G T A C G
                             3 2 1
                     x  G C A G A G A G
```

Shift by 4 ($bmGs[5] = bmBc[C] - 7 + 5$)

Fifth attempt:

```
y  G C A T C G C A G A G A G T A T A C A G T A C G
                                                2 1
                             x  G C A G A G A G
```

Shift by 7 ($bmGs[6]$)

The Galil algorithm performs 17 text character comparisons on the example.

5 References

- GALIL, Z., 1979, On improving the worst case running time of the Boyer–Moore string searching algorithm, *Communications of the ACM.* 22(9):505–508.

- LECROQ, T., 2000, *Quelques aspects de l'algorithmique du texte*, Habilitation thesis, Université de Rouen, France.

- SMYTH, W. F., 2003, *Computing Patterns in Strings*, Pearson Addison Wesley.

- STEPHEN, G.A., 1994, *String Searching Algorithms*, World Scientific.

CHAPTER 16

SMYTH ALGORITHM

1 Main Features

- improvement of the Galil algorithm;
- preprocessing phase in $O(m + \sigma)$ time and space complexity;
- searching phase in $O(n)$ time complexity;
- $4n$ text character comparisons in the worst case;
- $O(n/m)$ best performance.

2 Description

The prefix memorization technique that is used by the Galil algorithm (see chapter 15) only when an occurrence of the pattern has been found during the previous attempt can actually be applied each time that a prefix of the pattern is aligned with the suffix of the pattern recognized during the previous attempt. More formally, if the length of the shift is given by $bmGs[i]$ and such that $bmGs[i] > i$ then a prefix is memorized.

3 The C code

The functions preBmBc and preBmGs are given chapter 14.

At each attempt, the variable ell memorizes the left bound of the variable i.

```
void SMYTH(String x, int m, String y, int n) {
   int i, j, ell, bmGs[XSIZE], bmBc[ASIZE];

   /* Preprocessing */
   preBmGs(x, m, bmGs);
   preBmBc(x, m, bmBc);

   /* Searching */
   j = 0;
```

```
        ell = 0;
        while (j <= n - m) {
           for (i = m - 1; i >= 0 && x[i] == y[i + j]; --i);
           if (i < 0) {
              OUTPUT(j);
              j += bmGs[0];
              ell = m - bmGs[0];
           }
           else
              if (bmGs[i] >= bmBc[y[i + j]] - m + 1 + i) {
                 j += bmGs[i];
                 if (i < bmGs[i])
                    ell = m - bmGs[i];
                 else
                    ell = 0;
              }
              else {
                 j += bmBc[y[i + j]] - m + 1 + i;
                 ell = 0;
              }
        }
}
```

4 The example

c	A	C	G	T
$bmBc[c]$	1	6	2	8

i	0	1	2	3	4	5	6	7
$x[i]$	G	C	A	G	A	G	A	G
$suff[i]$	1	0	0	2	0	4	0	8
$bmGs[i]$	7	7	7	2	7	4	7	1

Searching phase

First attempt:

```
y  G C A T C G C A G A G A G T A T A C A G T A C G
                   1
x  G C A G A G A G
```

Shift by 1 ($bmGs[7] = bmBc[A] - 7 + 7$).

Second attempt:

```
y  G C A T C G C A G A G A G T A T A C A G T A C G
             3 2 1
x  G C A G A G A G
```
Shift by 4 ($bmGs[5] = bmBc[C] - 7 + 5$)

Third attempt:

```
y  G C A T C G C A G A G A G T A T A C A G T A C G
           8 7 6 5 4 3 2 1
         x G C A G A G A G
```
Shift by 7 ($bmGs[0]$)

Fourth attempt:

```
y  G C A T C G C A G A G A G T A T A C A G T A C G
                         3 2 1
                  x  G C A G A G A G
```
Shift by 4 ($bmGs[5] = bmBc[C] - 7 + 5$)

Fifth attempt:

```
y  G C A T C G C A G A G A G T A T A C A G T A C G
                                       2 1
                      x  G C A G A G A G
```
Shift by 7 ($bmGs[6]$)

The Smyth algorithm performs 17 text character comparisons on the example.

5 References

- LECROQ, T., 2000, *Quelques aspects de l'algorithmique du texte*, Habilitation thesis, Université de Rouen, France.

- SMYTH, W. F., 2003, *Computing Patterns in Strings*, Pearson Addison Wesley.

CHAPTER 17

TURBO-BM ALGORITHM

1 Main Features

- variant of the Boyer-Moore algorithm;

- no extra preprocessing needed with respect to the Boyer-Moore algorithm;

- constant extra space needed with respect to the Boyer-Moore algorithm;

- preprocessing phase in $O(m + \sigma)$ time and space complexity;

- searching phase in $O(n)$ time complexity;

- $2n$ text character comparisons in the worst case.

2 Description

The Turbo-BM algorithm is an amelioration of the Boyer-Moore algorithm (see chapter 14). It needs no extra preprocessing and requires only a constant extra space with respect to the original Boyer-Moore algorithm. It consists in remembering the factor of the text that matched a suffix of the pattern during the last attempt (and only if a good-suffix shift was performed).

This technique presents two advantages:

- it is possible to jump over this factor;

- it can enable to perform a **turbo-shift**.

A turbo-shift can occur if during the current attempt the suffix of the pattern that matches the text is shorter than the one remembered from the preceding attempt. In this case let us call u the remembered factor and v the suffix matched during the current attempt such that uzv is a suffix of x. Let a and b be the characters that cause the mismatch during the current attempt in the pattern and the text respectively. Then av is a suffix of x, and thus of u since $|v| < |u|$. The two characters a and b occur at distance p

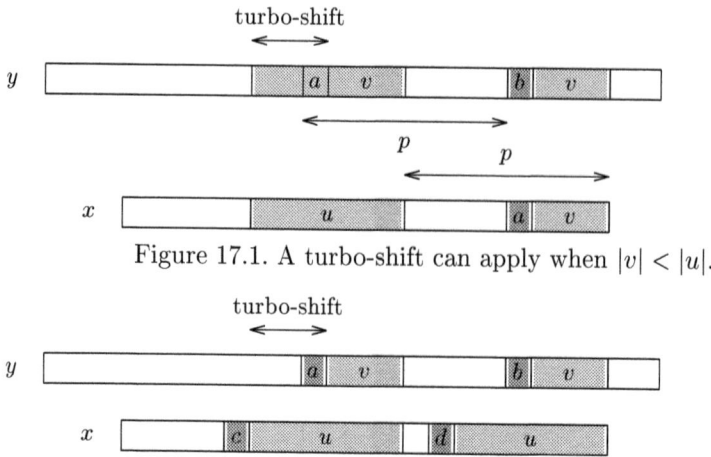

Figure 17.1. A turbo-shift can apply when $|v| < |u|$.

Figure 17.2. $c \neq d$ so they cannot be aligned with the same character in v.

in the text, and the suffix of x of length $|uzv|$ has a period of length $p = |zv|$ since u is a border of uzv, thus it cannot overlap both occurrences of two different characters a and b, at distance p, in the text. The smallest shift possible has length $|u| - |v|$, which is called a turbo-shift (see figure 17.1).

Still in the case where $|v| < |u|$ if the length of the bad-character shift is larger than the length of the good-suffix shift and the length of the turbo-shift then the length of the actual shift must be greater or equal to $|u| + 1$. Indeed (see figure 17.2), in this case the two characters c and d are different since we assumed that the previous shift was a good-suffix shift. Then a shift greater than the turbo-shift but smaller than $|u| + 1$ would align c and d with a same character in v. Thus if this case the length of the actual shift must be at least equal to $|u| + 1$.

The preprocessing phase can be performed in $O(m + \sigma)$ time and space complexity. The searching phase is in $O(n)$ time complexity. The number of text character comparisons performed by the Turbo-BM algorithm is bounded by $2n$.

3 The C code

The functions `preBmBc` and `preBmGs` are given chapter 14.

In the TBM function, the variable u memorizes the length of the suffix matched during the previous attempt and the variable v memorizes the length of the suffix matched during the current attempt.

17. TURBO-BM ALGORITHM

```
void TBM(String x, int m, String y, int n) {
   int bcShift, i, j, shift, u, v, turboShift,
       bmGs[XSIZE], bmBc[ASIZE];

   /* Preprocessing */
   preBmGs(x, m, bmGs);
   preBmBc(x, m, bmBc);

   /* Searching */
   j = u = 0;
   shift = m;
   while (j <= n - m) {
      i = m - 1;
      while (i >= 0 && x[i] == y[i + j]) {
         --i;
         if (u != 0 && i == m - 1 - shift)
            i -= u;
      }
      if (i < 0) {
         OUTPUT(j);
         shift = bmGs[0];
         u = m - shift;
      }
      else {
         v = m - 1 - i;
         turboShift = u - v;
         bcShift = bmBc[y[i + j]] - m + 1 + i;
         shift = MAX(turboShift, bcShift);
         shift = MAX(shift, bmGs[i]);
         if (shift == bmGs[i])
            u = MIN(m - shift, v);
         else {
           if (turboShift < bcShift)
              shift = MAX(shift, u + 1);
           u = 0;
         }
      }
      j += shift;
   }
}
```

4 The example

a	A	C	G	T
$bmBc[a]$	1	6	2	8

i	0	1	2	3	4	5	6	7
$x[i]$	G	C	A	G	A	G	A	G
$suff[i]$	1	0	0	2	0	4	0	8
$bmGs[i]$	7	7	7	2	7	4	7	1

Searching phase

First attempt:

```
y  G C A T C G C A G A G A G T A T A C A G T A C G
                   1
x  G C A G A G A G
```

Shift by 1 ($bmGs[7] = bmBc[A] - 7 + 7$)

Second attempt:

```
y  G C A T C G C A G A G A G T A T A C A G T A C G
             3 2 1
x    G C A G A G A G
```

Shift by 4 ($bmGs[5] = bmBc[C] - 7 + 5$)

Third attempt:

```
y  G C A T C G C A G A G A G T A T A C A G T A C G
           6 5     4 3 2 1
x          G C A G A G A G
```

Shift by 7 ($bmGs[0]$)

Fourth attempt:

```
y  G C A T C G C A G A G A G T A T A C A G T A C G
                                   3 2 1
x                        G C A G A G A G
```

Shift by 4 ($bmGs[5] = bmBc[C] - 7 + 5$)

Fifth attempt:

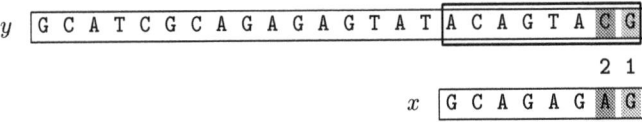

Shift by 7 ($bmGs[6]$)

The Turbo-BM algorithm performs 15 text character comparisons on the example.

5 References

- CROCHEMORE, M., 1997, Off-line serial exact string searching, in *Pattern Matching Algorithms*, A. Apostolico and Z. Galil eds., Chapter 1, pp 1–53, Oxford University Press.

- CROCHEMORE, M., CZUMAJ, A., GĄSIENIEC, L., JAROMINEK, S., LECROQ, T., PLANDOWSKI, W., RYTTER, W., 1992, Deux méthodes pour accélérer l'algorithme de Boyer-Moore, in *Théorie des Automates et Applications*, Actes des 2e Journées Franco-Belges, D. Krob ed., Rouen, France, 1991, pp 45–63, PUR 176, Rouen, France.

- CROCHEMORE, M., CZUMAJ, A., GĄSIENIEC, L., JAROMINEK, S., LECROQ, T., PLANDOWSKI, W., RYTTER, W., 1994, Speeding up two string matching algorithms, *Algorithmica* **12**(4/5):247–267.

- CROCHEMORE, M., HANCART, C., LECROQ, T., 2001, *Algorithmique du texte*, Vuibert.

- CROCHEMORE, M., RYTTER, W., 1994, *Text Algorithms*, Oxford University Press.

- LECROQ, T., 1992, *Recherches de mot*, Thèse de doctorat de l'Université d'Orléans, France.

- LECROQ, T., 1995, Experimental results on string matching algorithms, *Software – Practice & Experience* **25**(7):727-765.

- LECROQ, T., 2000, *Quelques aspects de l'algorithmique du texte*, Habilitation thesis, Université de Rouen, France.

- SMYTH, W. F., 2003, *Computing Patterns in Strings*, Pearson Addison Wesley.

CHAPTER 18

APOSTOLICO-GIANCARLO ALGORITHM

1 Main Features

- variant of the Boyer-Moore algorithm;
- preprocessing phase in $O(m + \sigma)$ time and space complexity;
- searching phase in $O(n)$ time complexity;
- $\frac{3}{2}n$ comparisons in the worst case.

2 Description

The Boyer-Moore algorithm (see chapter 14) is difficult to analyze because after each attempt it forgets all the characters it has already matched. Apostolico and Giancarlo designed an algorithm which remembers the length of the longest suffix of the pattern ending at the right position of the window at the end of each attempt. These information are stored in a table skip. Let us assume that during an attempt at a left position less than j the algorithm has matched a suffix of x of length k ending at position $i+j$ with $0 < i < m$ then skip$[i+j]$ is equal to k. Let suff$[i]$, for $0 \leq i < m$ be equal to the length of the longest suffix of x ending at the position i in x (see chapter 14). During the attempt at position j, if the algorithm compares successfully the factor of the text $y[i+j+1 .. j+m-1]$ then four cases arise:

Case 1: $k >$ suff$[i]$ and suff$[i] = i+1$. It means that an occurrence of x is found at position j and skip$[j+m-1]$ is set to m (see figure 18.1). A shift of length per(x) is performed.

Case 2: $k >$ suff$[i]$ and suff$[i] \leq i$. It means that a mismatch occurs between characters $x[i-$suff$[i]]$ and $y[i+j-$suff$[i]]$ and skip$[j+m-1]$ is set to $m-1-i+$suff$[i]$ (see figure 18.2). A shift is performed using bmBc$[y[i+j-$suff$[i]]]$ and bmGs$[i-$suff$[i]+1]$.

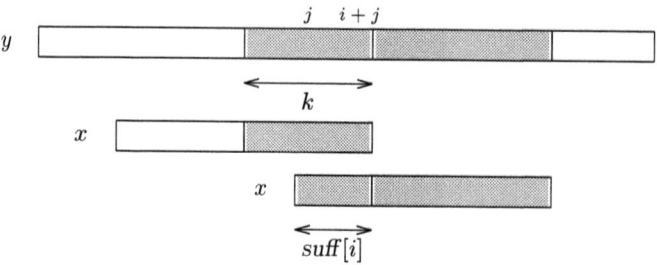

Figure 18.1. Case 1, $k > \text{suff}[i]$ and $\text{suff}[i] = i + 1$, an occurrence of x is found.

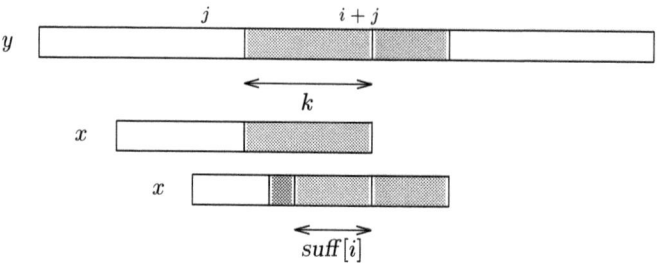

Figure 18.2. Case 2, $k > \text{suff}[i]$ and $\text{suff}[i] \leq i$, a mismatch occurs between $y[i + j - \text{suff}[i]]$ and $x[i - \text{suff}[i]]$.

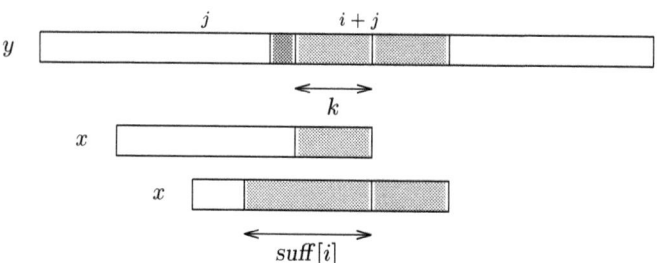

Figure 18.3. Case 3, $k < \text{suff}[i]$ a mismatch occurs between $y[i + j - k]$ and $x[i - k]$.

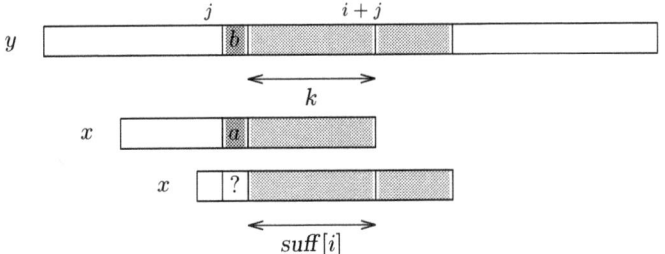

Figure 18.4. Case 4, $k = \mathit{suff}[i]$ and $a \neq b$.

Case 3: $k < \mathit{suff}[i]$. It means that a mismatch occurs between characters $x[i-k]$ and $y[i+j-k]$ and $\mathit{skip}[j+m-1]$ is set to $m-1-i+k$ (see figure 18.3). A shift is performed using $\mathit{bmBc}[y[i+j-k]]$ and $\mathit{bmGs}[i-k+1]$.

Case 4: $k = \mathit{suff}[i]$. This is the only case where a "jump" has to be done over the text factor $y[i+j-k+1..i+j]$ in order to resume the comparisons between the characters $y[i+j-k]$ and $x[i-k]$ (see figure 18.4).

In each case the only information which is needed is the length of the longest suffix of x ending at position i on x.

The Apostolico-Giancarlo algorithm use two data structures:

- a table skip which is updated at the end of each attempt j in the following way:

$$\mathit{skip}[j+m-1] = \max\{\ k \mid x[m-k..m-1] = y[j+m-k..j+m-1]\}$$

- the table suff used during the computation of the table bmGs:

$$\text{for } 1 \leq i < m, \mathit{suff}[i] = \max\{k \mid x[i-k+1..i] = x[m-k, m-1]\}$$

The complexity in space and time of the preprocessing phase of the Apostolico-Giancarlo algorithm is the same than for the Boyer-Moore algorithm: $O(m + \sigma)$.

During the search phase only the last m informations of the table skip are needed at each attempt so the size of the table skip can be reduced to $O(m)$. The Apostolico-Giancarlo algorithm performs in the worst case at most $\frac{3}{2}n$ text character comparisons.

3 The C code

The functions preBmBc and preBmGs are given chapter 14. It is enough to add the table *suff* as a parameter to the function preBmGs to get the correct values in the function AG.

```c
void AG(String x, int m, String y, int n) {
   int i, j, k, s, shift,
       bmGs[XSIZE], skip[XSIZE], suff[XSIZE], bmBc[ASIZE];

   /* Preprocessing */
   preBmGs(x, m, bmGs, suff);
   preBmBc(x, m, bmBc);
   memset(skip, 0, m*sizeof(int));

   /* Searching */
   j = 0;
   while (j <= n - m) {
      i = m - 1;
      while (i >= 0) {
         k = skip[i];
         s = suff[i];
         if (k > 0)
            if (k > s) {
               if (i + 1 == s)
                  i = (-1);
               else
                  i -= s;
               break;
            }
            else {
               i -= k;
               if (k < s)
                  break;
            }
         else {
            if (x[i] == y[i + j])
               --i;
            else
               break;
         }
      }
}
```

```
        if (i < 0) {
           OUTPUT(j);
           skip[m - 1] = m;
           shift = bmGs[0];
        }
        else {
           skip[m - 1] = m - 1 - i;
           shift = MAX(bmGs[i], bmBc[y[i + j]] - m + 1 + i);
        }
        j += shift;
        memcpy(skip, skip + shift, (m - shift)*sizeof(int));
        memset(skip + m - shift, 0, shift*sizeof(int));
     }
}
```

4 The example

a	A	C	G	T
$bmBc[a]$	1	6	2	8

i	0	1	2	3	4	5	6	7
$x[i]$	G	C	A	G	A	G	A	G
$suff[i]$	1	0	0	2	0	4	0	8
$bmGs[i]$	7	7	7	2	7	4	7	1

Searching phase

First attempt:

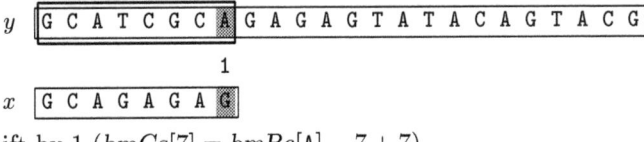

Shift by 1 ($bmGs[7] = bmBc[\text{A}] - 7 + 7$)

Second attempt:

Shift by 4 ($bmGs[5] = bmBc[\text{C}] - 7 + 5$)

Third attempt:

```
y  G C A T C G C A G A G A G T A T A C A G T A C G
           6 5       4 3 2 1
        x  G C A G A G A G
```

Shift by 7 ($bmGs[0]$)

Fourth attempt:

```
y  G C A T C G C A G A G A G T A T A C A G T A C G
                               3 2 1
                    x  G C A G A G A G
```

Shift by 4 ($bmGs[5] = bmBc[C] - 7 + 5$)

Fifth attempt:

```
y  G C A T C G C A G A G A G T A T A C A G T A C G
                                           2 1
                        x  G C A G A G A G
```

Shift by 7 ($bmGs[6]$)

The Apostolico-Giancarlo algorithm performs 15 text character comparisons on the example.

5 References

- APOSTOLICO, A., GIANCARLO, R., 1986, The Boyer-Moore-Galil string searching strategies revisited, *SIAM Journal on Computing*, **15**(1):98–105.

- CROCHEMORE, M., HANCART, C., LECROQ, T., 2001, *Algorithmique du texte*, Vuibert.

- CROCHEMORE, M., HANCART, C., LECROQ, T., A unifying look at the Apostolico–Giancarlo string matching algorithm, 2003, *Journal of Discrete Algorithms* **1**(1) 37–52.

- CROCHEMORE, M., LECROQ, T., 1997, Tight bounds on the complexity of the Apostolico-Giancarlo algorithm, *Information Processing Letters* **63**(4):195–203.

- CROCHEMORE, M., RYTTER, W., 1994, *Text Algorithms*, Oxford University Press.

- GUSFIELD, D., 1997, *Algorithms on strings, trees, and sequences: Computer Science and Computational Biology*, Cambridge University Press.

- LECROQ, T., 1992, *Recherches de mot*, Thèse de doctorat de l'Université d'Orléans, France.

- LECROQ, T., 1995, Experimental results on string matching algorithms, *Software – Practice & Experience* **25**(7):727-765.

- LECROQ, T., 2000, *Quelques aspects de l'algorithmique du texte*, Habilitation thesis, Université de Rouen, France.

CHAPTER 19

REVERSE COLUSSI ALGORITHM

1 Main features

- refinement of the Boyer-Moore algorithm;
- partitions the set of pattern positions into two disjoint subsets;
- preprocessing phase in $O(m^2)$ time and $O(m \times \sigma)$ space complexity;
- searching phase in $O(n)$ time complexity;
- $2n$ text character comparisons in the worst case.

2 Description

The character comparisons are done using a specific order given by a table h.

For each integer i such that $0 \leq i \leq m$ we define two disjoint sets:

$Pos(i) = \{k \mid 0 \leq k \leq i \text{ and } x[i] = x[i-k]\}$

and

$Neg(i) = \{k \mid 0 \leq k \leq i \text{ and } x[i] \neq x[i-k]\}$.

For $1 \leq k \leq m$, let $hmin[k]$ be the minimum integer ℓ such that $\ell \geq k-1$ and $k \notin Neg(i)$ for all i such that $\ell < i \leq m-1$.

For $0 \leq \ell \leq m-1$, let $kmin[\ell]$ be the minimum integer k such that $hmin[k] = \ell \geq k$ if any such k exists and $kmin[\ell] = 0$ otherwise.

For $0 \leq \ell \leq m-1$, let $rmin[\ell]$ be the minimum integer k such that $r > \ell$ and $hmin[r] = r - 1$.

The value of $h[0]$ is set to $m-1$.

After that we choose, all the indexes $h[1], \ldots, h[d]$, in increasing order of $kmin[\ell]$, such that $kmin[h[i]] \neq 0$ and we set $rcGs[i]$ to $kmin[h[i]]$ for $1 \leq i \leq d$.

Then we choose the indexes $h[d+1], \ldots, h[m-1]$ in increasing order and we set $rcGs[i]$ to $rmin[h[i]]$ for $d < i < m$.

The value of $rcGs[m]$ is set to the period of x.
The table $rcBc$ is defined as follows:

$$rcBc[a, s] = \min\{k \mid \begin{array}{l} (k = m \text{ or } x[m - k - 1] = a) \text{ and} \\ (k > m - s - 1 \text{ or} \\ x[m - k - s - 1] = x[m - s - 1]) \} \end{array}$$

To compute the table $rcBc$ we define: for each $c \in \Sigma$, $locc[c]$ is the index of the rightmost occurrence of c in $x[0..m-2]$ ($locc[c]$ is set to -1 if c does not occur in $x[0..m-2]$).

A table $link$ is used to link downward all the occurrences of each pattern character.

The preprocessing phase can be performed in $O(m^2)$ time and $O(m \times \sigma)$ space complexity. The searching phase is in $O(n)$ time complexity.

3 The C code

```
void preRc(String x, int m, int h[],
           int rcBc[ASIZE][XSIZE], int rcGs[]) {
  Character a;
  int i, j, k, q, r, s,
      hmin[XSIZE], kmin[XSIZE], link[XSIZE],
      locc[ASIZE], rmin[XSIZE];

  /* Computation of link and locc */
  for (a = 0; a < ASIZE; ++a)
     locc[a] = -1;
  link[0] = -1;
  for (i = 0; i < m - 1; ++i) {
     link[i + 1] = locc[x[i]];
     locc[x[i]] = i;
  }

  /* Computation of rcBc */
  for (a = 0; a < ASIZE; ++a)
     for (s = 1; s <= m; ++s) {
        i = locc[a];
        j = link[m - s];
        while (i - j != s && j >= 0)
           if (i - j > s)
              i = link[i + 1];
           else
              j = link[j + 1];
```

19. REVERSE COLUSSI ALGORITHM

```
         while (i - j > s)
            i = link[i + 1];
         rcBc[a][s] = m - i - 1;
   }

/* Computation of hmin */
k = 1;
i = m - 1;
while (k <= m) {
   while (i - k >= 0 && x[i - k] == x[i])
      --i;
   hmin[k] = i;
   q = k + 1;
   while (hmin[q - k] - (q - k) > i) {
      hmin[q] = hmin[q - k];
      ++q;
   }
   i += (q - k);
   k = q;
   if (i == m)
      i = m - 1;
}

/* Computation of kmin */
memset(kmin, 0, m * sizeof(int));
for (k = m; k > 0; --k)
   kmin[hmin[k]] = k;

/* Computation of rmin */
for (i = m - 1; i >= 0; --i) {
   if (hmin[i + 1] == i)
      r = i + 1;
   rmin[i] = r;
}

/* Computation of rcGs */
i = 1;
for (k = 1; k <= m; ++k)
   if (hmin[k] != m - 1 && kmin[hmin[k]] == k) {
      h[i] = hmin[k];
      rcGs[i++] = k;
```

```
      }
   i = m-1;
   for (j = m - 2; j >= 0; --j)
      if (kmin[j] == 0) {
         h[i] = j;
         rcGs[i--] = rmin[j];
      }
   rcGs[m] = rmin[0];
}


void RC(String x, int m, String y, int n) {
   int i, j, s, rcBc[ASIZE][XSIZE], rcGs[XSIZE], h[XSIZE];

   /* Preprocessing */
   preRc(x, m, h, rcBc, rcGs);

   /* Searching */
   j = 0;
   s = m;
   while (j <= n - m) {
      while (j <= n - m && x[m - 1] != y[j + m - 1]) {
         s = rcBc[y[j + m - 1]][s];
         j += s;
      }
      for (i = 1; i < m && x[h[i]] == y[j + h[i]]; ++i);
      if (i >= m)
         OUTPUT(j);
      s = rcGs[i];
      j += s;
   }
}
```

4 The example

a	A	C	G	T
$locc[a]$	6	1	5	-1

19. REVERSE COLUSSI ALGORITHM

rcBc	1	2	3	4	5	6	7	8
A	8	5	5	3	3	3	1	1
C	8	6	6	6	6	6	6	6
G	2	2	2	4	4	2	2	2
T	8	8	8	8	8	8	8	8

i	0	1	2	3	4	5	6	7	8
$x[i]$	G	C	A	G	A	G	A	G	
$link[i]$	-1	-1	-1	-1	0	2	3	4	
$hmin[i]$	0	7	3	7	5	5	7	6	7
$kmin[i]$	0	0	0	2	0	4	7	1	0
$rmin[i]$	7	7	7	7	7	7	7	8	0
$rcGs[i]$	0	2	4	7	7	7	7	7	7
$h[i]$		3	5	6	0	1	2	4	

Searching phase

First attempt:

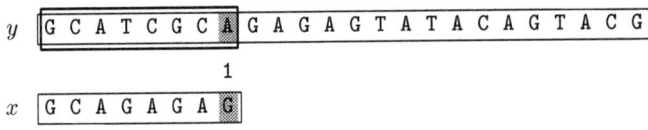

Shift by 1 ($rcBc[A][8]$)

Second attempt:

```
y  G C A T C G C A G A G A G T A T A C A G T A C G
         2       1
x    G C A G A G A G
```

Shift by 2 ($rcGs[1]$)

Third attempt:

```
y  G C A T C G C A G A G A G T A T A C A G T A C G
             2       1
x        G C A G A G A G
```

Shift by 2 ($rcGs[1]$)

Fourth attempt:

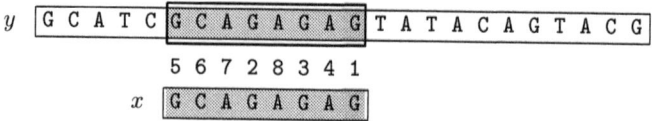

Shift by 7 ($rcGs[8]$)

Fifth attempt:

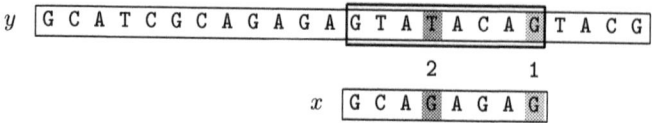

Shift by 2 ($rcGs[1]$)

Sixth attempt:

y | G C A T C G C A G A G A G T | A T A C A G T | A | C G |
$\quad\quad\quad\quad\quad\quad\quad\quad\quad\quad\quad\quad\quad\quad\quad\quad\quad\quad\quad\quad\quad\quad$ 1
x | G C A G A G A G |

Shift by 5 ($rcBc[A][2]$)

The Reverse-Colussi algorithm performs 16 text character comparisons on the example.

5 References

- COLUSSI, L., 1994, Fastest pattern matching in strings, *Journal of Algorithms*. **16**(2):163–189.

CHAPTER 20

HORSPOOL ALGORITHM

1 Main Features

- simplification of the Boyer-Moore algorithm;
- uses only the bad-character shift;
- easy to implement;
- preprocessing phase in $O(m + \sigma)$ time and $O(\sigma)$ space complexity;
- searching phase in $O(m \times n)$ time complexity;
- the average number of comparisons for one text character is between $1/\sigma$ and $2/(\sigma + 1)$.

2 Description

The bad-character shift which is used in the Boyer-Moore algorithm (see chapter 14) is not very efficient for small alphabets, but when the alphabet is large compared with the length of the pattern, as it is often the case with the ASCII table and ordinary searches made under a text editor, it becomes very useful. Using it alone produces a very efficient algorithm in practice. Horspool proposed to use only the bad-character shift of the rightmost character of the window to compute the shifts in the Boyer-Moore algorithm. The preprocessing phase is in $O(m + \sigma)$ time and $O(\sigma)$ space complexity.

The searching phase has a quadratic worst case but it can be proved that the average number of comparisons for one text character is between $1/\sigma$ and $2/(\sigma + 1)$.

3 The C code

The function preBmBc is given chapter 14.

```
void HORSPOOL(String x, int m, String y, int n) {
   int j, bmBc[ASIZE];
   Character c;
```

```
/* Preprocessing */
preBmBc(x, m, bmBc);

/* Searching */
j = 0;
while (j <= n - m) {
   c = y[j + m - 1];
   if (x[m - 1] == c && memcmp(x, y + j, m - 1) == 0)
      OUTPUT(j);
   j += bmBc[c];
}
}
```

4 The example

a	A	C	G	T
$bmBc[a]$	1	6	2	8

Searching phase

First attempt:

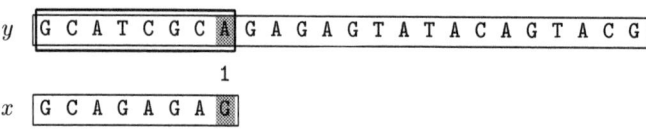

Shift by 1 ($bmBc[\text{A}]$)

Second attempt:

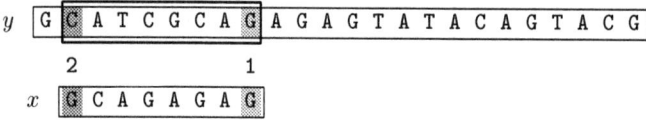

Shift by 2 ($bmBc[\text{G}]$)

Third attempt:

y |G C A T C G C A G A G|A G T A T A C A G T A C G
         2   1
    x |G C A G A G A G|

Shift by 2 ($bmBc[\text{G}]$)

Fourth attempt:

```
y  G C A T C G C A G A G A G T A T A C A G T A C G
           2 3 4 5 6 7 8 1
        x  G C A G A G A G
```
Shift by 2 ($bmBc[\texttt{G}]$)

Fifth attempt:

```
y  G C A T C G C A G A G A G T A T A C A G T A C G
                              1
              x  G C A G A G A G
```
Shift by 1 ($bmBc[\texttt{A}]$)

Sixth attempt:

```
y  G C A T C G C A G A G A G T A T A C A G T A C G
                               1
               x  G C A G A G A G
```
Shift by 8 ($bmBc[\texttt{T}]$)

Seventh attempt:

```
y  G C A T C G C A G A G A G T A T A C A G T A C G
                                      2           1
                       x  G C A G A G A G
```
Shift by 2 ($bmBc[\texttt{G}]$)

The Horspool algorithm performs 17 text character comparisons on the example.

5 References

- AHO, A.V., 1990, Algorithms for Finding Patterns in Strings, in *Handbook of Theoretical Computer Science, Volume A, Algorithms and complexity*, J. van Leeuwen ed., Chapter 5, pp 255–300, Elsevier, Amsterdam.

- BAEZA-YATES, R.A., RÉGNIER, M., 1992, Average running time of the Boyer-Moore-Horspool algorithm, *Theoretical Computer Science* **92**(1): 19–31.

- BEAUQUIER, D., BERSTEL, J., CHRÉTIENNE, P., 1992, Éléments d'algorithmique, Chapter 10, pp 337–377, Masson, Paris.

- CROCHEMORE, M., HANCART, C., 1999, Pattern Matching in Strings, in Algorithms and Theory of Computation Handbook, M.J. Atallah ed., Chapter 11, pp 11-1–11-28, CRC Press Inc., Boca Raton, FL.

- CROCHEMORE, M., HANCART, C., LECROQ, T., 2001, Algorithmique du texte, Vuibert.

- HANCART, C., 1993, Analyse exacte et en moyenne d'algorithmes de recherche d'un motif dans un texte, Thèse de doctorat de l'Université de Paris 7, France.

- HORSPOOL, R.N., 1980, Practical fast searching in strings, Software – Practice & Experience, 10(6):501–506.

- LECROQ, T., 1995, Experimental results on string matching algorithms, Software – Practice & Experience 25(7):727-765.

- NAVARRO, G., RAFFINOT, M., 2002, Flexible Pattern Matching in Strings Practical on-line search algorithms for texts and biological sequences, Cambridge University Press.

- SMYTH, W. F., 2003, Computing Patterns in Strings, Pearson Addison Wesley.

- STEPHEN, G.A., 1994, String Searching Algorithms, World Scientific.

CHAPTER 21

FAST SEARCH ALGORITHM

1 Main Features

- hybrid between the Boyer-Moore algorithm and the Horspool algorithm;
- easy to implement;
- preprocessing phase in $O(m + \sigma)$ time and space complexity;
- searching phase in $O(m \times n)$ time complexity;
- very fast in practice.

2 Description

The Fast Search algorithm is an hybrid between the Boyer-Moore algorithm (see chapter 14 and the Horspool algorithm (see chapter 20). It uses the the bad-character shift whenever a mismatch occurs when comparing the last character of the pattern $x[m-1]$. When a mismatch occurs with another character of the pattern or when an occurrence of x is found, it uses the good-sufffix shift.

This is due to the following properties:

- $bmBc[c] \geq bmGs[m-1]$ for $c \in \Sigma$;

- during an attempt at a left position j (the window contains $y[j \mathinner{.\,.} j + m - 1]$ and $0 \leq j \leq n - m$) if a mismatch occurs between $x[i]$ and $y[i+j]$ (with $0 \leq i \leq m - 2$) then $bmGs[i] \leq bmBc[y[i+j]]$.

The preprocessing phase is in $O(m + \sigma)$ time and space complexity.

The searching phase has a quadratic worst case but it is very fast in practice.

3 The C code

The functions preBmBc and preBmGs are given chapter 14.

```
void FASTSEARCH(String x, int m, String y, int n) {
   int i, j, bmGs[XSIZE], bmBc[ASIZE];

   /* Preprocessing */
   preBmBc(x, m, bmBc);
   preBmGs(x, m, bmGs);
   memset(y + n, x[m - 1], m);
   bmBc[x[m - 1]] = 0;

   /* Searching */
   for (j = m - 1; bmBc[y[j]] > 0; j += bmBc[y[j]]);
   while (j < n) {
      for (i = m - 2;
           i >= 0 && x[i] == y[i + j - m + 1];
           --i);
      if (i < 0) {
         OUTPUT(j - m + 1);
         j += bmGs[0];
      }
      else
         j += bmGs[i];
      while (bmBc[y[j]] > 0)
         j += bmBc[y[j]];
   }
}
```

4 The example

a	A	C	G	T
$bmBc[a]$	1	6	0	8

i	0	1	2	3	4	5	6	7
$x[i]$	G	C	A	G	A	G	A	G
$bmGs[i]$	7	7	7	2	7	4	7	1

Searching phase

First attempt:

```
y  G C A T C G C A G A G A G T A T A C A G T A C G
                   1
x  G C A G A G A G
```

Shift by 1 ($bmBc[\text{A}]$)

Second attempt:

```
y  G C A T C G C A G A G A G T A T A C A G T A C G
           3 2 1
    x  G C A G A G A G
```

Shift by 4 ($bmGs[5]$)

Third attempt:

```
y  G C A T C G C A G A G A G T A T A C A G T A C G
       8 7 6 5 4 3 2 1
        x  G C A G A G A G
```

Shift by 7 ($bmGs[0]$)

Fourth attempt:

```
y  G C A T C G C A G A G A G T A T A C A G T A C G
                          3 2 1
                x  G C A G A G A G
```

Shift by 4 ($bmGs[5]$)

Fifth attempt:

```
y  G C A T C G C A G A G A G T A T A C A G T A C G
                                      2 1
                    x  G C A G A G A G
```

Shift by 7 ($bmGs[6]$)

The Fast Search algorithm performs 17 text character comparisons on the example.

5 References

- CANTONE, D., FARO, S., 2003, Fast-Search: A New Efficient Variant of the Boyer-Moore String Matching Algorithm, in *Proceedings of the 2nd International Workshop on Experimental and Efficient Algorithms*, K. Jansen, M. Margraf, M. Mastrolilli and J. D. P. Rolim eds., Ascona, Switzerland, Lecture Notes in Computer Science 2647, pp 47–58, Springer-Verlag, Berlin.

CHAPTER 22

QUICK SEARCH ALGORITHM

1 Main Features

- simplification of the Boyer-Moore algorithm;
- uses only the bad-character shift;
- easy to implement;
- preprocessing phase in $O(m + \sigma)$ time and $O(\sigma)$ space complexity;
- searching phase in $O(m \times n)$ time complexity;
- very fast in practice for short patterns and large alphabets.

2 Description

The Quick Search algorithm uses only the bad-character shift table (see chapter 14). After an attempt where the window is positioned on the text factor $y[j \mathinner{.\,.} j + m - 1]$, the length of the shift is at least equal to one. So, the character $y[j+m]$ is necessarily involved in the next attempt, and thus can be used for the bad-character shift of the current attempt. The bad-character shift of the present algorithm is slightly modified to take into account the last character of x as follows: for $c \in \Sigma$

$$qsBc[c] = \begin{cases} \min\{i \mid 0 \leq i < m \text{ and } x[m-1-i] = c\} & \text{if } c \text{ occurs in } x, \\ m & \text{otherwise}. \end{cases}$$

The preprocessing phase is in $O(m + \sigma)$ time and $O(\sigma)$ space complexity.

During the searching phase the comparisons between pattern and text characters during each attempt can be done in any order. The searching phase has a quadratic worst case time complexity but it has a good practical behavior.

3 The C code

```
void preQsBc(String x, int m, int qsBc[]) {
   int i;

   for (i = 0; i < ASIZE; ++i)
      qsBc[i] = m + 1;
   for (i = 0; i < m; ++i)
      qsBc[x[i]] = m - i;
}


void QS(String x, int m, String y, int n) {
   int j, qsBc[ASIZE];

   /* Preprocessing */
   preQsBc(x, m, qsBc);

   /* Searching */
   j = 0;
   while (j <= n - m) {
      if (memcmp(x, y + j, m) == 0)
         OUTPUT(j);
      j += qsBc[y[j + m]];              /* shift */
   }
}
```

4 The example

a	A	C	G	T
$qsBc[a]$	2	7	1	9

Searching phase

First attempt:

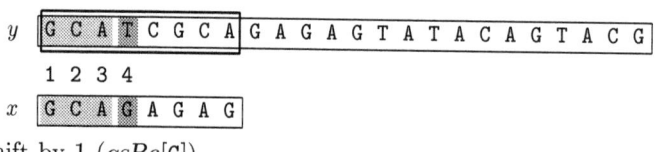

Shift by 1 ($qsBc[\text{G}]$)

Second attempt:

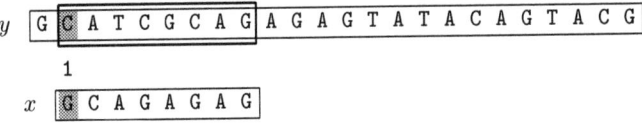

Shift by 2 ($qsBc[\text{A}]$)

Third attempt:

```
y  G C A T C G C A G A G T A T A C A G T A C G
         1
x      G C A G A G A G
```

Shift by 2 ($qsBc[\text{A}]$)

Fourth attempt:

```
y  G C A T C G C A G A G A G T A T A C A G T A C G
             1 2 3 4 5 6 7 8
x          G C A G A G A G
```

Shift by 9 ($qsBc[\text{T}]$)

Fifth attempt:

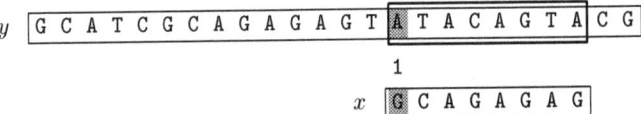

Shift by 7 ($qsBc[\text{C}]$)

The Quick Search algorithm performs 15 text character comparisons and it inspects 5 more text characters in order to compute the shifts, on the example.

5 References

- CROCHEMORE, M., LECROQ, T., 1996, Pattern matching and text compression algorithms, in *CRC Computer Science and Engineering Handbook*, A.B. Tucker Jr ed., Chapter 8, pp 162–202, CRC Press Inc., Boca Raton, FL.

- LECROQ, T., 1995, Experimental results on string matching algorithms, *Software – Practice & Experience* **25**(7):727-765.

- SMYTH, W. F., 2003, *Computing Patterns in Strings*, Pearson Addison Wesley.

- STEPHEN, G.A., 1994, *String Searching Algorithms*, World Scientific.

- SUNDAY, D.M., 1990, A very fast substring search algorithm, *Communications of the ACM* **33**(8):132–142.

CHAPTER 23

TURBO SEARCH ALGORITHM

1 Main Features

- improvement of the Quick Search algorithm;
- uses an extra shift whenever the last character of the pattern is matched;
- easy to implement;
- very fast in practice.

2 Description

The Turbo Search algorithm is an improvement of the Quick Search algorithm (see chapter 22). At each attempt, it always tries to match the last character of the pattern first. Whenever it finds it an extra shift is used, otherwise the shifts are computed as in the Quick Search algorithm. The $m-1$ first characters of the pattern can be compared in any order.

More formally the variable *shift* is defined as follows:

$$\text{shift} = \min\{\ell \mid 1 < \ell \leq m \text{ and } x[m-\ell] = x[m-1]\} - 2$$

if $x[m-1]$ occurs in $x[0..m-2]$ and *shift* is equal to $m-1$ otherwise.

Then at an attempt at a left position j in the text (the window is positioned at $y[j..j+m-1]$), if $x[m-1] = y[j+m-1]$ the shift following the attempt is given by $shift + qsBc[y[m+shift]]$.

This algorithm has a quadratic worst-case time complexity but a very good practical behavior.

3 The C code

The function `preQsBc` is given chapter 22.

```
void TS(String x, int m, String y, int n) {
   int j, k, shift, qsBc[ASIZE];
```

```
/* Preprocessing */
preQsBc(x, m, qsBc);
for (shift = 0;
     shift < m - 1 && x[m - 2 - shift] != x[m - 1];
     ++shift);
memset(y + n, x[m - 1], m);

/* Searching */
j = 0;
while (j < n) {
   while (x[m - 1] != y[j + m - 1])
      j += qsBc[y[j + m]];
   if (memcmp(x, y + j, m - 1) == 0 && j < n)
      OUTPUT(j);
   j += shift + qsBc[y[j + m + shift]];      /* shift */
}
}
```

4 The example

a	A	C	G	T
$qsBc[a]$	2	7	1	9

$shift = 1$

Searching phase

First attempt:

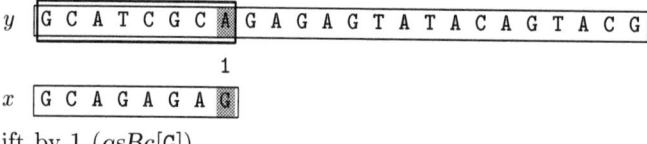

Shift by 1 ($qsBc[\text{G}]$)

Second attempt:

y | G | C | A | T | C | G | C | A | G | A | G | A | G | T | A | T | A | C | A | G | T | A | C | G |
     | | 2 | | | | | | | 1 | | | | | | | | | | | | | | | |
x | G | C | A | G | A | G | A | G |

Shift by 2 ($1 + qsBc[\text{G}]$)

Third attempt:

```
y  G C A T C G C A G A G A G T A T A C A G T A C G
       2           1
x      G C A G A G A G
```
Shift by 2 $(1 + qsBc[\text{G}])$

Fourth attempt:

```
y  G C A T C G C A G A G A G T A T A C A G T A C G
         2 3 4 5 6 7 8 1
x        G C A G A G A G
```
Shift by 3 $(1 + qsBc[\text{A}])$

Fifth attempt:

```
y  G C A T C G C A G A G A G T A T A C A G T A C G
                         1
x              G C A G A G A G
```
Shift by 2 $(qsBc[\text{A}])$

Sixth attempt:

```
y  G C A T C G C A G A G A G T A T A C A G T A C G
                             1
x                G C A G A G A G
```
Shift by 2 $(qsBc[\text{A}])$

Seventh attempt:

```
y  G C A T C G C A G A G A G T A T A C A G T A C G
                     2 3           1
x                    G C A G A G A G
```
Shift by 3 $(1 + qsBc[\text{A}])$

Eighth attempt:

```
y  G C A T C G C A G A G A G T A T A C A G T A C G
                                                1
                              x  G C A G A G A G
```

Shift by 1 ($qsBc[\texttt{G}]$)

Ninth attempt:

```
y  G C A T C G C A G A G A G T A T A C A G T A C G
                                      2           1
                                x  G C A G A G A G
```

Shift by 2 ($1 + qsBc[\texttt{G}]$)

The Turbo Search algorithm performs 21 text character comparisons and it inspects 9 more text characters in order to compute the shifts, on the example.

5 References

- TAMM, M., 1997, Blitzfindig: Texte schnell durchsuchen mit T-Search, c't **8**:292.

CHAPTER 24

TUNED BOYER-MOORE ALGORITHM

1 Main Features

- simplification of the Boyer-Moore algorithm;
- easy to implement;
- very fast in practice.

2 Description

The Tuned Boyer-Moore is a implementation of a simplified version of the Boyer-Moore algorithm which is very fast in practice. The most costly part of a string matching algorithm is to check whether the character of the pattern match the character of the window. To avoid doing this part too often, it is possible to unrolled several shifts before actually comparing the characters. The algorithm used the bad-character shift function to find $x[m-1]$ in y and keep on shifting until finding it, doing blindly three shifts in a row. This required to save the value of $bmBc[x[m-1]]$ in a variable *shift* and then to set $bmBc[x[m-1]]$ to 0. This required also to add m occurrences of $x[m-1]$ at the end of y. When $x[m-1]$ is found the $m-1$ other characters of the window are checked and a shift of length *shift* is applied.

The comparisons between pattern and text characters during each attempt can be done in any order. This algorithm has a quadratic worst-case time complexity but a very good practical behavior.

Depending on the underlying alphabet, the number of shifts that are performed in a row can be tuned to fit the user needs.

3 The C code

The function preBmBc is given chapter 14.

```
void TUNEDBM(String x, int m, String y, int n) {
   int j, k, shift, bmBc[ASIZE];

   /* Preprocessing */
   preBmBc(x, m, bmBc);
   shift = bmBc[x[m - 1]];
   bmBc[x[m - 1]] = 0;
   memset(y + n, x[m - 1], m);

   /* Searching */
   j = 0;
   while (j < n) {
      k = bmBc[y[j + m -1]];
      while (k !=  0) {
         j += k; k = bmBc[y[j + m -1]];
         j += k; k = bmBc[y[j + m -1]];
         j += k; k = bmBc[y[j + m -1]];
      }
      if (memcmp(x, y + j, m - 1) == 0 && j < n)
         OUTPUT(j);
      j += shift;                             /* shift */
   }
}
```

4 The example

a	A	C	G	T
$bmBc[a]$	1	6	0	8

$shift = 2$

Searching phase

First attempt:

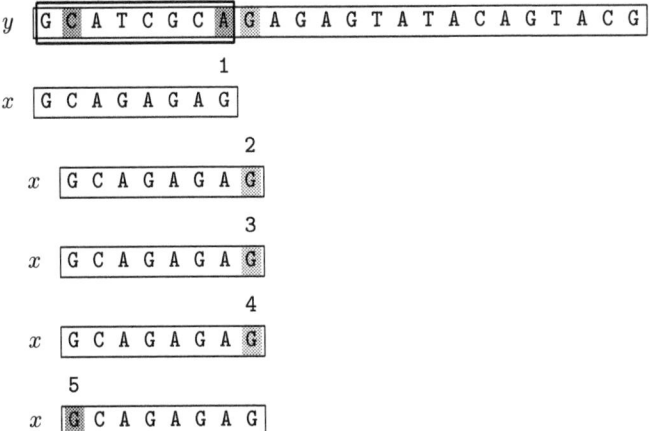

Shift by 2 (*shift*)

Second attempt:

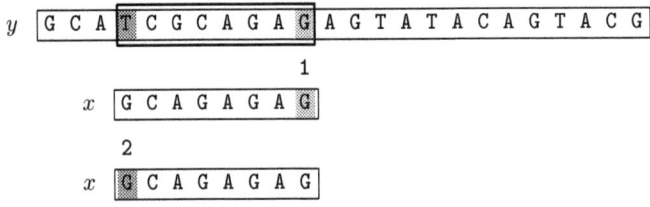

Shift by 2 (*shift*)

Third attempt:

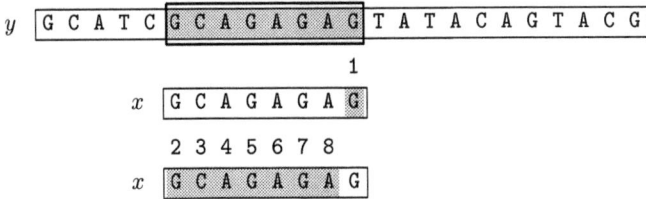

Shift by 2 (*shift*)

Fourth attempt:

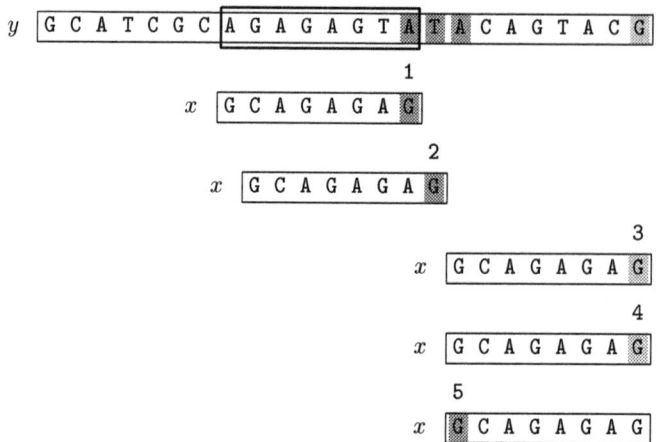

Shift by 2 (*shift*)

The Tuned Boyer-Moore algorithm performs 20 text character comparisons on the example.

5 References

- HUME, A., SUNDAY, D.M., 1991, Fast string searching, *Software – Practice & Experience* **21**(11):1221–1248.

- STEPHEN, G.A., 1994, *String Searching Algorithms*, World Scientific.

CHAPTER 25

ZHU-TAKAOKA ALGORITHM

1 Main features

- variant of the Boyer-Moore algorithm;

- uses two consecutive text characters to compute the bad-character shift;

- preprocessing phase in $O(m + \sigma^2)$ time and space complexity;

- searching phase in $O(m \times n)$ time complexity.

2 Description

Zhu and Takaoka designed an algorithm which performs the shift by considering the bad-character shift (see chapter 14) for two consecutive text characters. During the searching phase the comparisons are performed from right to left and when the window is positioned on the text factor $y[j\mathrel{..}j+m-1]$ and a mismatch occurs between $x[m-k]$ and $y[j+m-k]$ while $x[m-k+1\mathrel{..}m-1] = y[j+m-k+1\mathrel{..}j+m-1]$ the shift is performed with the bad-character shift for text characters $y[j+m-2]$ and $y[j+m-1]$. The good-suffix shift table is also used to compute the shifts.

The preprocessing phase of the algorithm consists in computing for each pair of characters (a,b) with $a,b \in \Sigma$ the rightmost occurrence of ab in $x[0\mathrel{..}m-2]$.

For $a, b \in \Sigma$:

$$ztBc[a,b] = k \Leftrightarrow \begin{cases} k < m-2 & \text{and } x[m-k..m-k+1] = ab \\ & \text{and } ab \text{ does not occur} \\ & \text{in } x[m-k+2..m-2], \\ \text{or} \\ k = m-1 & x[0] = b \text{ and } ab \text{ does not occur} \\ & \text{in } x[0..m-2], \\ \text{or} \\ k = m & x[0] \neq b \text{ and } ab \text{ does not occur} \\ & \text{in } x[0..m-2]. \end{cases}$$

It also consists in computing the table $bmGs$ (see chapter 14). The preprocessing phase is in $O(m + \sigma^2)$ time and space complexity.

The searching phase has a quadratic worst case.

3 The C code

The function preBmGs is given chapter 14.

```
void preZtBc(String x, int m, int ztBc[ASIZE][ASIZE]) {
   int i, j;

   for (i = 0; i < ASIZE; ++i)
      for (j = 0; j < ASIZE; ++j)
         ztBc[i][j] = m;
   for (i = 0; i < ASIZE; ++i)
      ztBc[i][x[0]] = m - 1;
   for (i = 1; i < m - 1; ++i)
      ztBc[x[i - 1]][x[i]] = m - 1 - i;
}

void ZT(String x, int m, String y, int n) {
   int i, j, ztBc[ASIZE][ASIZE], bmGs[XSIZE];

   /* Preprocessing */
   preZtBc(x, m, ztBc);
   preBmGs(x, m, bmGs);
```

```
/* Searching */
j = 0;
while (j <= n - m) {
    i = m - 1;
    while (i < m && x[i] == y[i + j])
        --i;
    if (i < 0) {
        OUTPUT(j);
        j += bmGs[0];
    }
    else
        j += MAX(bmGs[i],
                ztBc[y[j + m - 2]][y[j + m - 1]]);
}
}
```

4 The example

$ztBc$	A	C	G	T
A	8	8	2	8
C	5	8	7	8
G	1	6	7	8
T	8	8	7	8

i	0	1	2	3	4	5	6	7
$x[i]$	G	C	A	G	A	G	A	G
$bmGs[i]$	7	7	7	2	7	4	7	1

Searching phase

First attempt:

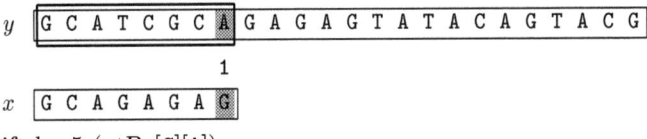

Shift by 5 ($ztBc[C][A]$)

Second attempt:

y | G C A T C G C A G A G A G T A T A C A G T A C G

8 7 6 5 4 3 2 1

x | G C A G A G A G

Shift by 7 ($bmGs[0]$)

Third attempt:

```
y  G C A T C G C A G A G A G T A T A C A G T A C G
                              3 2 1
              x              G C A G A G A G
```

Shift by 4 ($bmGs[6]$)

Fourth attempt:

```
y  G C A T C G C A G A G A G T A T A C A G T A C G
                                            2 1
                      x              G C A G A G A G
```

Shift by 7 ($bmGs[7] = ztBc[C][G]$)

The Zhu-Takaoka algorithm performs 14 text character comparisons and it inspects one more text character in order to compute the first shift, on the example.

5 References

- ZHU, R.F., TAKAOKA, T., 1987, On improving the average case of the Boyer-Moore string matching algorithm, *Journal of Information Processing* **10**(3):173–177.

CHAPTER 26

BERRY-RAVINDRAN ALGORITHM

1 Main features

- hybrid of the Quick Search and Zhu-Takaoka algorithms;
- preprocessing phase in $O(m + \sigma^2)$ space and time complexity;
- searching phase in $O(m \times n)$ time complexity.

2 Description

Berry and Ravindran designed an algorithm which performs the shifts by considering the bad-character shift (see chapter 14) for the two consecutive text characters immediately to the right of the window.

The preprocessing phase of the algorithm consists in computing for each pair of characters (a, b) with $a, b \in \Sigma$ the rightmost occurrence of ab in axb. For $a, b \in \Sigma$

$$brBc[a, b] = \min \begin{cases} 1 & \text{if } x[m-1] = a, \\ m-i+1 & \text{if } x[i]x[i+1] = ab, \\ m+1 & \text{if } x[0] = b, \\ m+2 & \text{otherwise}. \end{cases}$$

The preprocessing phase is in $O(m + \sigma^2)$ space and time complexity.

After an attempt where the window is positioned on the text factor $y[j \mathinner{.\,.} j + m - 1]$ a shift of length $brBc[y[j + m], y[j + m + 1]]$ is performed. The text character $y[n]$ is equal to the null character and $y[n + 1]$ is set to this null character in order to be able to compute the last shifts of the algorithm.

The searching phase of the Berry-Ravindran algorithm has a $O(m \times n)$ time complexity.

3 The C code

```
void preBrBc(String x, int m, int brBc[ASIZE][ASIZE]) {
   Character a, b;
   int i;

   for (a = 0; a < ASIZE; ++a)
      for (b = 0; b < ASIZE; ++b)
         brBc[a][b] = m + 2;
   for (a = 0; a < ASIZE; ++a)
      brBc[a][x[0]] = m + 1;
   for (i = 0; i < m - 1; ++i)
      brBc[x[i]][x[i + 1]] = m - i;
   for (a = 0; a < ASIZE; ++a)
      brBc[x[m - 1]][a] = 1;
}


void BR(String x, int m, String y, int n) {
   int j, brBc[ASIZE][ASIZE];

   /* Preprocessing */
   preBrBc(x, m, brBc);

   /* Searching */
   y[n + 1] = '\0';
   j = 0;
   while (j <= n - m) {
      if (memcmp(x, y + j, m) == 0)
         OUTPUT(j);
      j += brBc[y[j + m]][y[j + m + 1]];
   }
}
```

4 The example

brBc	A	C	G	T	*
A	10	10	2	10	10
C	7	10	9	10	10
G	1	1	1	1	1
T	10	10	9	10	10
*	10	10	9	10	10

The star (*) represents any character in $\Sigma \setminus \{A, C, G, T\}$.

Searching phase

First attempt:

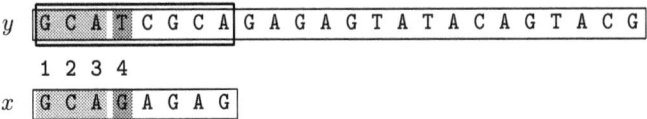

Shift by 1 ($brBc[\texttt{G}][\texttt{A}]$)

Second attempt:

```
y  G C A T C G C A G A G A G T A T A C A G T A C G
     1
x    G C A G A G A G
```

Shift by 2 ($brBc[\texttt{A}][\texttt{G}]$)

Third attempt:

```
y  G C A T C G C A G A G A G T A T A C A G T A C G
         1
x        G C A G A G A G
```

Shift by 2 ($brBc[\texttt{A}][\texttt{G}]$)

Fourth attempt:

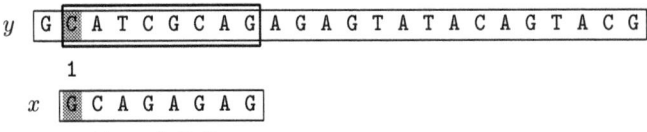

Shift by 10 ($brBc[\texttt{T}][\texttt{A}]$)

Fifth attempt:

```
y  G C A T C G C A G A G A G T A T A C A G T A C G
                                 1
x                                G C A G A G A G
```

Shift by 1 ($brBc[\texttt{G}][0]$)

Sixth attempt:

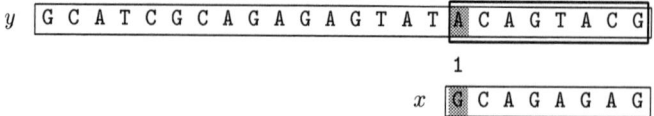

Shift by 10 ($brBc[0][0]$)

The Berry-Ravindran algorithm performs 16 text character comparisons and it inspects 12 more text characters in order to compute the shifts, on the example.

5 References

- BERRY, T., RAVINDRAN, S., 1999, A fast string matching algorithm and experimental results, in *Proceedings of the Prague Stringology Club Workshop'99*, J. Holub and M. Šimánek eds., Collaborative Report DC–99–05, Czech Technical University, Prague, Czech Republic, 1999, pp 16–26.

CHAPTER 27

SMITH ALGORITHM

1 Main features

- takes the maximum of the Horspool bad-character shift function and the Quick Search bad-character shift function;
- preprocessing phase in $O(m+\sigma)$ time and $O(\sigma)$ space complexity;
- searching phase in $O(m \times n)$ time complexity.

2 Description

Smith noticed that computing the shift with the text character just next the rightmost text character of the window gives sometimes shorter shift than using the rightmost text character of the window. He advised then to take the maximum between the two values.

The preprocessing phase of the Smith algorithm consists in computing the bad-character shift function (see chapter 14) and the Quick Search bad-character shift function (see chapter 22).

The preprocessing phase is in $O(m+\sigma)$ time and $O(\sigma)$ space complexity.

The searching phase of the Smith algorithm has a quadratic worst case time complexity.

3 The C code

The function preBmBc is given chapter 14 and the function preQsBc is given chapter 22.

```
void SMITH(String x, int m, String y, int n) {
   int j, bmBc[ASIZE], qsBc[ASIZE];

   /* Preprocessing */
   preBmBc(x, m, bmBc);
   preQsBc(x, m, qsBc);
```

```
/* Searching */
j = 0;
while (j <= n - m) {
   if (memcmp(x, y + j, m) == 0)
      OUTPUT(j);
   j += MAX(bmBc[y[j + m - 1]], qsBc[y[j + m]]);
}
}
```

4 The example

a	A	C	G	T
$bmBc[a]$	1	6	2	8
$qsBc[a]$	2	7	1	9

Searching phase

First attempt:

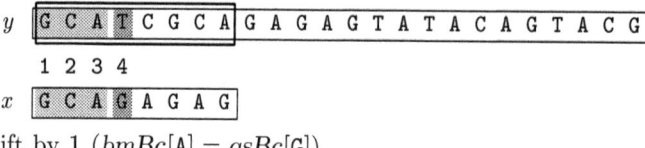

Shift by 1 ($bmBc[\text{A}] = qsBc[\text{G}]$)

Second attempt:

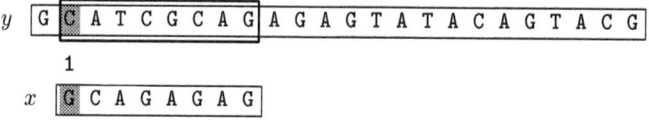

Shift by 2 ($bmBc[\text{G}] = qsBc[\text{A}]$)

Third attempt:

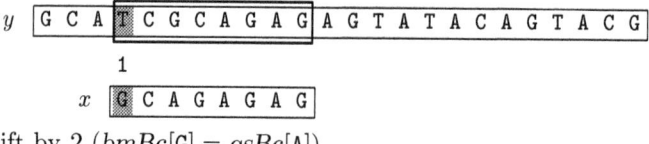

Shift by 2 ($bmBc[\text{G}] = qsBc[\text{A}]$)

Fourth attempt:

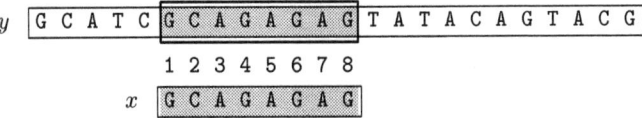

Shift by 9 ($qsBc[\text{T}]$)

Fifth attempt:

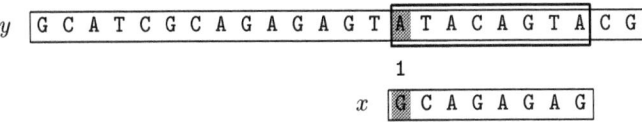

Shift by 7 ($qsBc[\text{C}]$)

The Smith algorithm performs 15 text character comparisons and it inspects 9 more text characters in order to compute the shifts, on the example.

5 References

- SMITH, P.D., 1991, Experiments with a very fast substring search algorithm, *Software – Practice & Experience* **21**(10):1065–1074.

CHAPTER 28

RAITA ALGORITHM

1 Main features

- first compares the last pattern character, then the first and finally the middle one before actually comparing the others;
- performs the shifts like the Horspool algorithm;
- preprocessing phase in $O(m + \sigma)$ time and $O(\sigma)$ space complexity;
- searching phase in $O(m \times n)$ time complexity.

2 Description

Raita designed an algorithm which at each attempt first compares the last character of the pattern with the rightmost text character of the window, then if they match it compares the first character of the pattern with the leftmost text character of the window, then if they match it compares the middle character of the pattern with the middle text character of the window. And finally if they match it actually compares the other characters from the second to the last but one, possibly comparing again the middle character.

Raita observed that its algorithm had a good behavior in practice when searching patterns in English texts and attributed these performance to the existence of character dependencies. Smith made some more experiments and concluded that this phenomenon may rather be due to compiler effects.

The preprocessing phase of the Raita algorithm consists in computing the bad-character shift function (see chapter 14). It can be done in $O(m + \sigma)$ time and $O(\sigma)$ space complexity.

The searching phase of the Raita algorithm has a quadratic worst case time complexity.

3 The C code

The function preBmBc is given chapter 14.

```
void RAITA(String x, int m, String y, int n) {
   int j, bmBc[ASIZE];
   Character c, firstCh, *secondCh, middleCh, lastCh;

   /* Preprocessing */
   preBmBc(x, m, bmBc);
   firstCh = x[0];
   secondCh = x + 1;
   middleCh = x[m/2];
   lastCh = x[m - 1];

   /* Searching */
   j = 0;
   while (j <= n - m) {
      c = y[j + m - 1];
      if (lastCh == c && middleCh == y[j + m/2] &&
          firstCh == y[j] &&
          memcmp(secondCh, y + j + 1, m - 2) == 0)
         OUTPUT(j);
      j += bmBc[c];
   }
}
```

4 The example

a	A	C	G	T
$bmBc[a]$	1	6	2	8

Searching phase

First attempt:

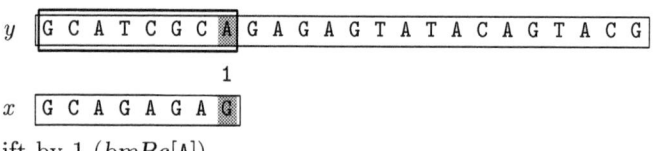

Shift by 1 ($bmBc[\text{A}]$)

Second attempt:

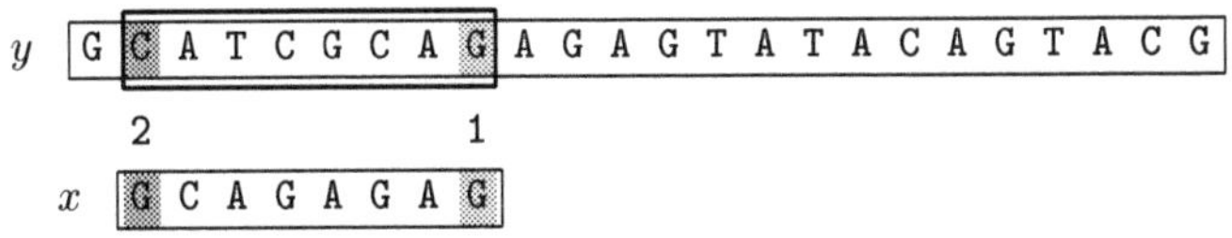

Shift by 2 ($bmBc$[G])

Third attempt:

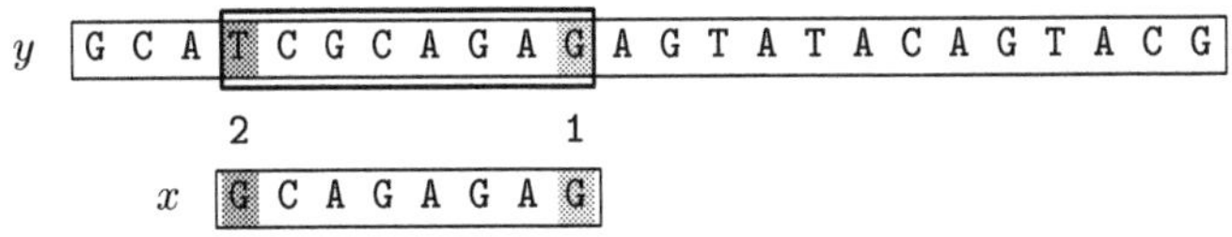

Shift by 2 ($bmBc$[G])

Fourth attempt:

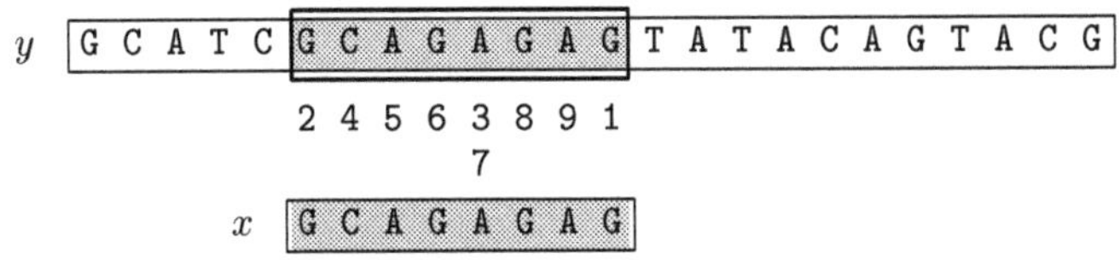

Shift by 2 ($bmBc$[G])

Fifth attempt:

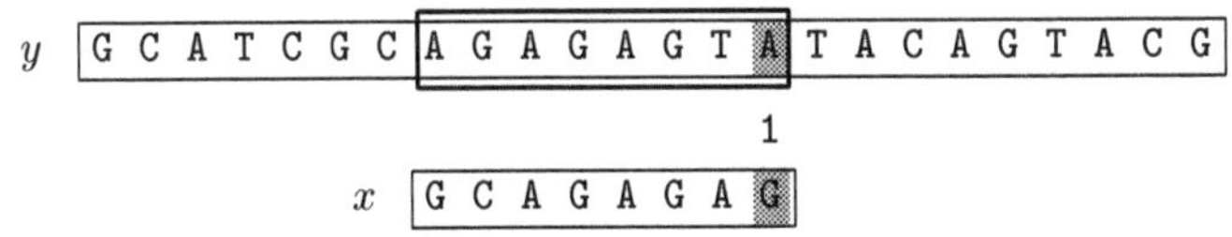

Shift by 1 ($bmBc$[A])

Sixth attempt:

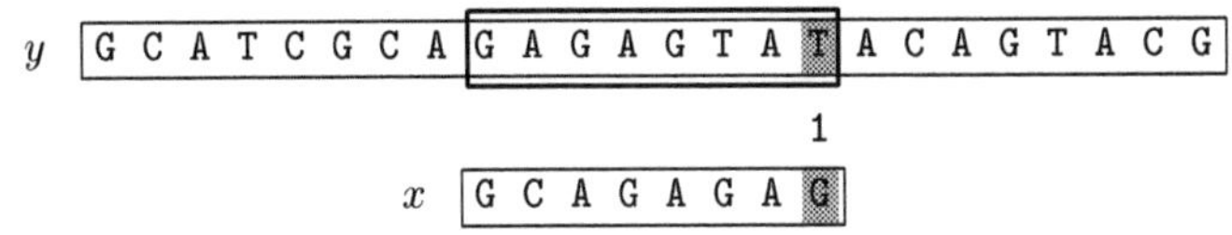

Shift by 8 ($bmBc$[T])

Seventh attempt:

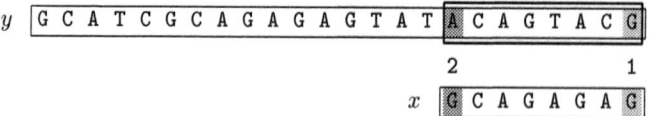

Shift by 2 ($bmBc[\text{G}]$)

The Raita algorithm performs 18 text character comparisons on the example.

5 References

- RAITA, T., 1992, Tuning the Boyer-Moore-Horspool string searching algorithm, *Software – Practice & Experience*, **22**(10):879–884.

- SMITH, P.D., 1994, On tuning the Boyer-Moore-Horspool string searching algorithms, *Software – Practice & Experience*, **24**(4):435–436.

CHAPTER 29

REVERSE FACTOR ALGORITHM

1 Main Features

- uses the suffix automaton of x^R;
- fast on practice for long patterns and small alphabets;
- preprocessing phase in $O(m)$ time and space complexity;
- searching phase in $O(m \times n)$ time complexity;
- optimal in the average.

2 Description

The Boyer-Moore (see chapter 14) type algorithms match some suffixes of the pattern but it is possible to match some prefixes of the pattern by scanning the character of the window from right to left and then improve the length of the shifts. This is made possible by the use of the smallest suffix automaton (also called DAWG for Directed Acyclic Word Graph) of the reverse pattern. The resulting algorithm is called the Reverse Factor algorithm.

The smallest suffix automaton of a word w is a Deterministic Finite Automaton $S(w) = (Q, q_0, T, E)$. The language accepted by $S(w)$ is $\mathcal{L}(S(w)) = \{u \in \Sigma^* \mid \exists v \in \Sigma^* \text{ such that } w = vu\}$. The preprocessing phase of the Reverse Factor algorithm consists in computing the smallest suffix automaton for the reverse pattern x^R. It is linear in time and space in the length of the pattern.

During the searching phase, the Reverse Factor algorithm parses the characters of the window from right to left with the automaton $S(x^R)$, starting with state q_0. It goes until there is no more transition defined for the current character of the window from the current state of the automaton. At this moment it is easy to know what is the length of the longest prefix of the pattern which has been matched: it corresponds to the length of the path taken in $S(x^R)$ from the start state q_0 to the last final state encountered.

Knowing the length of this longest prefix, it is trivial to compute the right shift to perform.

The Reverse Factor algorithm has a quadratic worst case time complexity but it is optimal in average. It performs $O(n \times (\log_\sigma m)/m)$ inspections of text characters on the average reaching the best bound shown by Yao in 1979.

3 The C code

All the functions to create and manipulate a data structure suitable for a suffix automaton are given section 5.

```
void buildSuffixAutomaton(String x, int m, Graph aut) {
   int i, art, init, last, p, q, r;
   Character c;

   init = getInitial(aut);
   art = newVertex(aut);
   setSuffixLink(aut, init, art);
   last = init;
   for (i = 0; i < m; ++i) {
      c = x[i];
      p = last;
      q = newVertex(aut);
      setLength(aut, q, getLength(aut, p) + 1);
      setPosition(aut, q, getPosition(aut, p) + 1);
      while (p != init &&
             getTarget(aut, p, c) == UNDEFINED) {
         setTarget(aut, p, c, q);
         setShift(aut, p, c, getPosition(aut, q) -
                             getPosition(aut, p) - 1);
         p = getSuffixLink(aut, p);
      }
      if (getTarget(aut, p, c) == UNDEFINED) {
         setTarget(aut, init, c, q);
         setShift(aut, init, c,
                  getPosition(aut, q) -
                  getPosition(aut, init) - 1);
         setSuffixLink(aut, q, init);
      }
      else
         if (getLength(aut, p) + 1 ==
             getLength(aut, getTarget(aut, p, c)))
```

```
            setSuffixLink(aut, q, getTarget(aut, p, c));
        else {
            r = newVertex(aut);
            copyVertex(aut, r, getTarget(aut, p, c));
            setLength(aut, r, getLength(aut, p) + 1);
            setSuffixLink(aut, getTarget(aut, p, c), r);
            setSuffixLink(aut, q, r);
            while (p != art &&
                    getLength(aut, getTarget(aut, p, c)) >=
                    getLength(aut, r)) {
                setShift(aut, p, c,
                        getPosition(aut,
                                getTarget(aut, p, c)) -
                        getPosition(aut, p) - 1);
                setTarget(aut, p, c, r);
                p = getSuffixLink(aut, p);
            }
        }
        last = q;
    }
    setTerminal(aut, last);
    while (last != init) {
        last = getSuffixLink(aut, last);
        setTerminal(aut, last);
    }
}


String reverse(String x, int m) {
    String xR;
    int i;

    xR = (String)malloc((m + 1)*sizeof(Character));
    for (i = 0; i < m; ++i)
        xR[i] = x[m - 1 - i];
    xR[m] = '\0';
    return(xR);
}
```

```
void RF(String x, int m, String y, int n) {
   int i, j, shift, period, init, state;
   Graph aut;
   String xR;

   /* Preprocessing */
   aut = newSuffixAutomaton(2*(m + 2), 2*(m + 2)*ASIZE);
   xR = reverse(x, m);
   buildSuffixAutomaton(xR, m, aut);
   init = getInitial(aut);
   period = m;

   /* Searching */
   j = 0;
   while (j <= n - m) {
      i = m - 1;
      state = init;
      shift = m;
      while (i + j >= 0 &&
             getTarget(aut, state, y[i + j]) !=
             UNDEFINED) {
         state = getTarget(aut, state, y[i + j]);
         if (isTerminal(aut, state)) {
            period = shift;
            shift = i;
         }
         --i;
      }
      if (i < 0) {
         OUTPUT(j);
         shift = period;
      }
      j += shift;
   }
}
```

The test i + j >= 0 in the inner loop of the searching phase of the function RF is only necessary during the first attempt, if x occurs at position 0 on y. Thus, in practice, to avoid testing at all the following attempts the first attempt could be distinguished from all the others.

4 The example

$\mathcal{L}(\mathcal{S}) = \{\text{GCAGAGAG, GCAGAGA, GCAGAG, GCAGA, GCAG, GCA, GC, G}, \varepsilon\}$

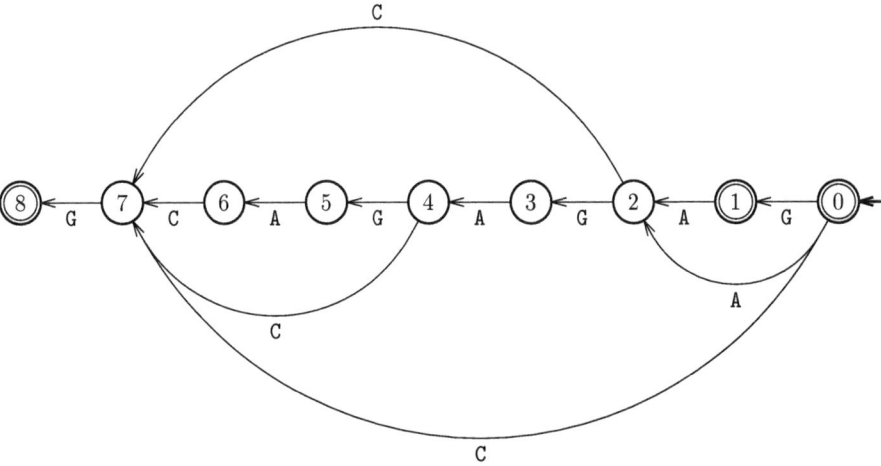

Searching phase

The initial state is 0.
First attempt:

y	G C A T **C G C A** G A G A G T A T A C A G T A C G
	* 8 7 2
x	G C A G **A G A G**

Shift by 5 (8-3)

Second attempt:

y	G C A T C **G C A G A G A G** T A T A C A G T A C G
	* 8 7 6 5 4 3 2 1
x	G C A G A G A G

Shift by 7 (8-1)

Third attempt:

y	G C A T C G C A G A G A G T A T **A C A G** T A C G
	* 7 2 1
x	G C A G **A G A G**

Shift by 7 (8-1)

The Reverse Factor algorithm performs 17 text character inspections on the example.

5 References

- BAEZA-YATES, R.A., NAVARRO G., RIBEIRO-NETO B., 1999, Indexing and Searching, in *Modern Information Retrieval*, Chapter 8, pp 191–228, Addison-Wesley.

- CROCHEMORE, M., CZUMAJ, A., GĄSIENIEC, L., JAROMINEK, S., LECROQ, T., PLANDOWSKI, W., RYTTER, W., 1992, Deux méthodes pour accélérer l'algorithme de Boyer-Moore, in *Théorie des Automates et Applications, Actes des 2e Journées Franco-Belges*, D. Krob ed., Rouen, France, 1991, pp 45–63, PUR 176, Rouen, France.

- CROCHEMORE, M., CZUMAJ, A., GĄSIENIEC, L., JAROMINEK, S., LECROQ, T., PLANDOWSKI, W., RYTTER, W., 1994, Speeding up two string matching algorithms, *Algorithmica* **12**(4/5):247–267.

- CROCHEMORE, M., RYTTER, W., 1994, *Text Algorithms*, Oxford University Press.

- LECROQ, T., 1992, A variation on the Boyer-Moore algorithm, *Theoretical Computer Science* **92**(1):119–144.

- LECROQ, T., 1992, *Recherches de mot*, Thèse de doctorat de l'Université d'Orléans, France.

- LECROQ, T., 1995, Experimental results on string matching algorithms, *Software – Practice & Experience* **25**(7):727-765.

- LECROQ, T., 2000, *Quelques aspects de l'algorithmique du texte*, Habilitation thesis, Université de Rouen, France.

- NAVARRO, G., RAFFINOT, M., 2002, *Flexible Pattern Matching in Strings Practical on-line search algorithms for texts and biological sequences*, Cambridge University Press.

- YAO, A.C., 1979, The complexity of pattern matching for a random string *SIAM Journal on Computing*, **8** (3):368–387.

CHAPTER 30

TURBO REVERSE FACTOR ALGORITHM

1 Main Features

- refinement of the Reverse Factor algorithm;
- preprocessing phase in $O(m)$ time and space complexity;
- searching phase in $O(n)$ time complexity;
- performs $2n$ text character inspections in the worst case;
- optimal in the average.

2 Description

It is possible to make the Reverse Factor algorithm (see chapter 29) linear. It is, in fact, enough to remember the prefix u of x matched during the last attempt. Then during the current attempt when reaching the right end of u, it is easy to show that it is sufficient to read again at most the rightmost half of u. This is made by the Turbo Reverse Factor algorithm.

If a word z is a factor of a word w we define $shift(z, w)$ the displacement of z in w to be the least integer $d > 0$ such that $w[m-d-|z|-1..m-d] = z$.

The general situation of the Turbo Reverse Factor algorithm is when a prefix u is found in the text during the last attempt and for the current attempt the algorithm tries to match the factor v of length $m - |u|$ in the text immediately at the right of u. If v is not a factor of x then the shift is computed as in the Reverse Factor algorithm. If v is a suffix of x then an occurrence of x has been found. If v is not a suffix but a factor of x then it is sufficient to scan again the $\min\{per(u), |u|/2\}$ rightmost characters of u. If u is periodic (i.e. $per(u) \leq |u|/2$) let z be the suffix of u of length $per(u)$. By definition of the period z is an acyclic word and then an overlap such as shown in figure 30.1 is impossible.

Thus z can only occur in u at distances multiple of $per(u)$ which implies that the smallest proper suffix of uv which is a prefix of x has a length equal to $|uv| - shift(zv, x) = m - shift(zv, x)$. Thus the length of the shift to perform is $shift(zv, x)$.

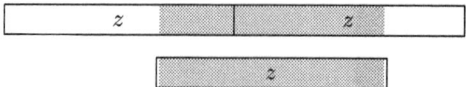

Figure 30.1. Impossible overlap if z is an acyclic word.

If u is not $(per(u) > |u|/2)$, it is obvious that x can not re-occur in the left part of u of length $per(u)$. It is then sufficient to scan the right part of u of length $|u| - per(u) < |u|/2$ to find a non defined transition in the automaton. The function shift is implemented directly in the automaton $S(x)$ without changing the complexity of its construction.

The preprocessing phase consists in building the suffix automaton of x^R. It can be done in $O(m)$ time complexity.

The searching phase is in $O(n)$ time complexity. The Turbo Reverse Factor performs at most $2n$ inspections of text characters and it is also optimal in average performing $O(n \times (\log_\sigma m)/m)$ inspections of text characters on the average reaching the best bound shown by Yao in 1979.

3 The C code

The function preMp is given chapter 6. The two functions reverse and buildSuffixAutomaton are given chapter 29. All the other functions to create and manipulate a data structure suitable for a suffix automaton are given section 5.

```
void TRF(String x, int m, String y, int n) {
   int period, i, j, shift, u, periodOfU, disp, init,
       state, mu, mpNext[XSIZE + 1];
   String xR;
   Graph aut;

   /* Preprocessing */
   aut = newSuffixAutomaton(2*(m + 2), 2*(m + 2)*ASIZE);
   xR = reverse(x, m);
   buildSuffixAutomaton(xR, m, aut);
   init = getInitial(aut);
   preMp(x, m, mpNext);
   period = m - mpNext[m];
   i = 0;
   shift = m;
```

30. TURBO REVERSE FACTOR ALGORITHM

```
/* Searching */
j = 0;
while (j <= n - m) {
   i = m - 1;
   state = init;
   u = m - 1 - shift;
   periodOfU = (shift != m ?
                m - shift - mpNext[m - shift] : 0);
   shift = m;
   disp = 0;
   while (i > u &&
          getTarget(aut, state, y[i + j]) !=
          UNDEFINED) {
      disp += getShift(aut, state, y[i + j]);
      state = getTarget(aut, state, y[i + j]);
      if (isTerminal(aut, state))
         shift = i;
      --i;
   }
   if (i <= u)
      if (disp == 0) {
         OUTPUT(j);
         shift = period;
      }
      else {
         mu = (u + 1)/2;
         if (periodOfU <= mu) {
            u -= periodOfU;
            while (i > u &&
                   getTarget(aut, state, y[i + j]) !=
                   UNDEFINED) {
               disp += getShift(aut, state, y[i + j]);
               state = getTarget(aut, state, y[i + j]);
               if (isTerminal(aut, state))
                  shift = i;
               --i;
            }
            if (i <= u)
               shift = disp;
         }
         else {
```

```
                u = u - mu - 1;
                while (i > u &&
                        getTarget(aut, state, y[i + j]) !=
                        UNDEFINED) {
                    disp += getShift(aut, state, y[i + j]);
                    state = getTarget(aut, state, y[i + j]);
                    if (isTerminal(aut, state))
                        shift = i;
                    --i;
                }
            }
        }
        j += shift;
    }
}
```

4 The example

$\mathcal{L}(\mathcal{S}) = \{\text{GCAGAGAG}, \text{GCAGAGA}, \text{GCAGAG}, \text{GCAGA}, \text{GCAG}, \text{GCA}, \text{GC}, \text{G}, \varepsilon\}$

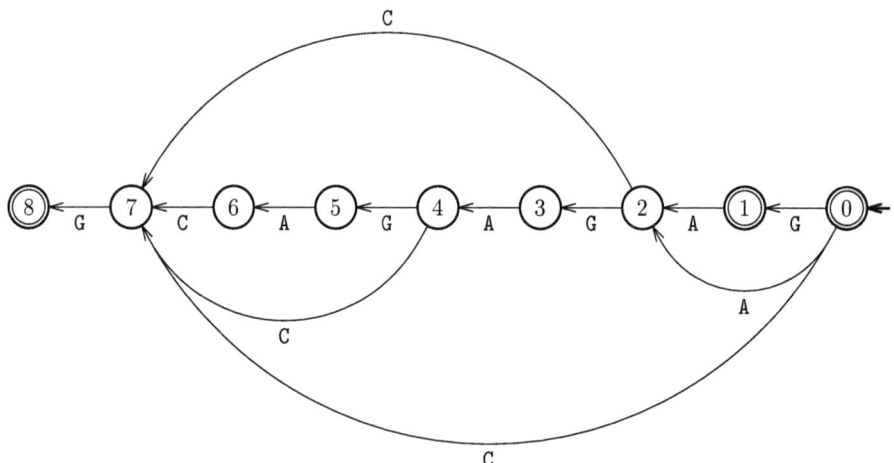

30. TURBO REVERSE FACTOR ALGORITHM

shift	A	C	G	T
0	1	6	0	
1	0			
2		4	0	
3	0			
4		2	0	
5	0			
6		0		
7			0	
8				

Searching phase

The initial state is 0.
First attempt:

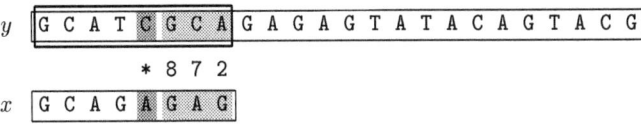

Shift by 5 (8-3)

Second attempt:

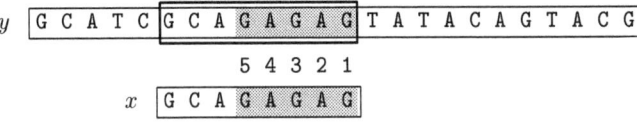

Shift by 7 (8-1)

Third attempt:

y G C A T C G C A G A G A G T A T A C A G T A C G
                                      * 7 2 1
                            x G C A G A G A G

Shift by 7 (8-1)

The Turbo Reverse Factor algorithm performs 13 text character inspections on the example.

5 References

- CROCHEMORE, M., 1997, Off-line serial exact string searching, in *Pattern Matching Algorithms*, A. Apostolico and Z. Galil eds., Chapter

1, pp 1–53, Oxford University Press.

- CROCHEMORE, M., CZUMAJ, A., GĄSIENIEC, L., JAROMINEK, S., LECROQ, T., PLANDOWSKI, W., RYTTER, W., 1992, Deux méthodes pour accélérer l'algorithme de Boyer-Moore, in *Théorie des Automates et Applications, Actes des 2e Journées Franco-Belges*, D. Krob ed., Rouen, France, 1991, pp 45–63, PUR 176, Rouen, France.

- CROCHEMORE, M., CZUMAJ, A., GĄSIENIEC, L., JAROMINEK, S., LECROQ, T., PLANDOWSKI, W., RYTTER, W., 1994, Speeding up two string matching algorithms, *Algorithmica* **12**(4/5):247–267.

- CROCHEMORE, M., RYTTER, W., 1994, *Text Algorithms*, Oxford University Press.

- LECROQ, T., 1992, *Recherches de mot*, Thèse de doctorat de l'Université d'Orléans, France.

- LECROQ, T., 1995, Experimental results on string matching algorithms, *Software – Practice & Experience* **25**(7):727-765.

- LECROQ, T., 2000, *Quelques aspects de l'algorithmique du texte*, Habilitation thesis, Université de Rouen, France.

- YAO, A.C., 1979, The complexity of pattern matching for a random string *SIAM Journal on Computing*, **8** (3):368–387.

CHAPTER 31

BACKWARD SUFFIX ORACLE MATCHING ALGORITHM

1 Main Features

- version of the Reverse Factor algorithm using the suffix oracle of x^R instead of the suffix automaton of x^R;
- fast in practice for very long patterns and small alphabets;
- preprocessing phase in $O(m)$ time and space complexity;
- searching phase in $O(m \times n)$ time complexity;
- optimal in the average.

2 Description

The Boyer-Moore (see chapter 14) type algorithms match some suffixes of the pattern but it is possible to match some prefixes of the pattern by scanning the character of the window from right to left and then improve the length of the shifts. This is make possible by the use of the suffix oracle of the reverse pattern. This data structure is a very compact automaton which recognizes at least all the suffixes of a word and slightly more other words. The string matching algorithm using the oracle of the reverse pattern is called the Backward Oracle Matching algorithm. The suffix oracle of a word w is a Deterministic Finite Automaton $\mathcal{O}(w) = (Q, q_0, T, E)$. The language accepted by $\mathcal{O}(w)$ is such that $\{u \in \Sigma^* \mid \exists v \in \Sigma^* \text{ such that } w = vu\} \subseteq \mathcal{L}(\mathcal{O}(w))$. The preprocessing phase of the Backward Oracle Matching algorithm consists in computing the suffix oracle for the reverse pattern x^R. Despite the fact that it is able to recognize words that are not factor of the pattern, the suffix oracle can be used to do string matching since the only word of length greater or equal m which is recognized by the oracle is the reverse pattern itself. The computation of the oracle is linear in time and space in the length of the pattern. During the searching phase the Backward Oracle Matching algorithm parses the characters of the window from right to left with the automaton $\mathcal{O}(x^R)$ starting with state q_0. It goes until there is no more transition defined for the current character. At this moment the length of the longest prefix of the pattern which is a suffix of the scanned part of the text is less than the length of the path taken in $\mathcal{O}(x^R)$ from the start state q_0 and the last final state encountered. Knowing this length, it is trivial to compute the length of the shift to perform.

The Backward Suffix Oracle Matching algorithm has a quadratic worst case time complexity but it is optimal in average. On the average it performs $O(n(\log_\sigma m)/m)$ inspections of text characters reaching the best bound shown by Yao in 1979.

3 The C code

Only the external transitions of the oracle are stored in link lists (one per state). The labels of these transitions and all the other transitions are not stored but computed from the word x. The description of a linked list List can be found section 5.

```
#define FALSE    0
#define TRUE     1

int getTransition(String x, int p, List L[], Character c) {
   List cell;

   if (p > 0 && x[p - 1] == c)
      return(p - 1);
   else {
      cell = L[p];
      while (cell != NULL)
         if (x[cell->element] == c)
            return(cell->element);
         else
            cell = cell->next;
      return(UNDEFINED);
   }
}


void setTransition(int p, int q, List L[]) {
   List cell;

   cell = (List)malloc(sizeof(struct _cell));
   if (cell == NULL)
```

```
      error("BSOM/setTransition");
   cell->element = q;
   cell->next = L[p];
   L[p] = cell;
}


void oracle(String x, int m, char T[], List L[]) {
   int i, p, q;
   int S[XSIZE + 1];
   Character c;

   S[m] = m + 1;
   for (i = m; i > 0; --i) {
      c = x[i - 1];
      p = S[i];
      while (p <= m &&
             (q = getTransition(x, p, L, c)) ==
             UNDEFINED) {
         setTransition(p, i - 1, L);
         p = S[p];
      }
      S[i - 1] = (p == m + 1 ? m : q);
   }
   p = 0;
   while (p <= m) {
      T[p] = TRUE;
      p = S[p];
   }
}


void BSOM(String x, int m, String y, int n) {
   char T[XSIZE + 1];
   List L[XSIZE + 1];
   int i, j, p, period, q, shift;

   /* Preprocessing */
   memset(L, NULL, (m + 1)*sizeof(List));
   memset(T, FALSE, (m + 1)*sizeof(char));
   oracle(x, m, T, L);
```

```
/* Searching */
j = 0;
while (j <= n - m) {
    i = m - 1;
    p = m;
    shift = m;
    while (i + j >= 0 &&
            (q = getTransition(x, p, L, y[i + j])) !=
            UNDEFINED) {
        p = q;
        if (T[p] == TRUE) {
            period = shift;
            shift = i;
        }
        --i;
    }
    if (i < 0) {
        OUTPUT(j);
        shift = period;
    }
    j += shift;
}
```

The test i + j >= 0 in the inner loop of the searching phase of the function BSOM is only necessary during the first attempt if x occurs at position 0 on y. Thus to avoid testing at all the following attempts the first attempt could be distinguished from all the others.

4 The example

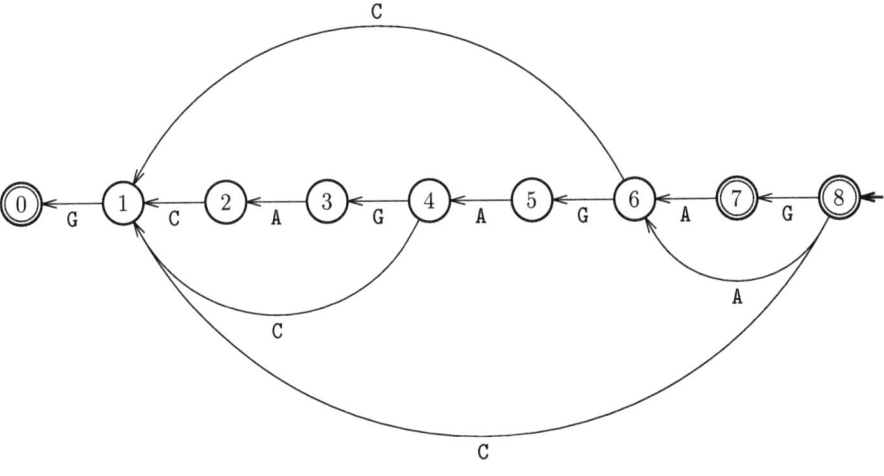

i	0	1	2	3	4	5	6	7	8
$x[i]$	G	C	A	G	A	G	A	G	
$S[i]$	7	8	4	5	6	7	8	8	9
$L[i]$	∅	∅	∅	∅	(1)	∅	(1)	∅	(1,6)

Searching phase

The initial state is 8.
First attempt:

```
y  G C A T C G C A G A G A G T A T A C A G T A C G
         *  0 1 6
x  G C A G A G A G
```
Shift by 5 (8-3)

Second attempt:

```
y  G C A T C G C A G A G A G T A T A C A G T A C G
           *  0 1 2 3 4 5 6 7
         x  G C A G A G A G
```
Shift by 7 (8-1)

Third attempt:

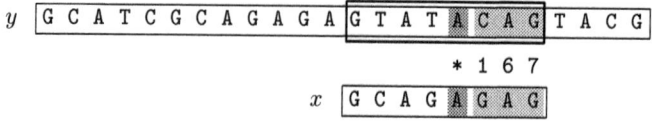

Shift by 7 (8-1)

The Backward Suffix Oracle Matching algorithm performs 17 text character inspections on the example.

5 References

- ALLAUZEN, C., CROCHEMORE, M., RAFFINOT M., 1999, Factor oracle: a new structure for pattern matching, in *Proceedings of SOFSEM'99, Theory and Practice of Informatics*, J. Pavelka, G. Tel and M. Bartosek eds., Milovy, Czech Republic, Lecture Notes in Computer Science 1725, pp 291–306, Springer-Verlag, Berlin.

- ALLAUZEN, C., CROCHEMORE, M., RAFFINOT M., 2001, Efficient Experimental String Matching by Weak Factor Recognition, in *Proceedings of the 12th Annual Symposium on Combinatorial Pattern Matching*, A. Amir and G. M. Landau eds., Jerusalem, Israel, Lecture Notes in Computer Science 2089, pp 51–72, Springer-Verlag, Berlin.

- NAVARRO, G., RAFFINOT, M., 2002, *Flexible Pattern Matching in Strings Practical on-line search algorithms for texts and biological sequences*, Cambridge University Press.

CHAPTER 32

BACKWARD NONDETERMINISTIC DAWG MATCHING ALGORITHM

1 Main Features

- variant of the Reverse Factor algorithm;
- uses bit-parallelism simulation of the suffix automaton of x^R;
- efficient if the pattern length is no longer than the memory-word size of the machine;

2 Description

The Backward Nondeterministic Dawg Matching algorithm uses a table B which, for each character c, stores a bit mask. The i-th bit of $B[c]$ is set to 1 if and only if $x[i] = c$.

The search state is kept in a word $d = d[m-1]..d[0]$, where the pattern length m is less than or equal to the machine word size.

During an attempt at left position j (the window is positioned on the text factor $y[j..j+m-1])$, the text character are scanned from right to left. After scanning $y[j+m-k]$, the bit $d[i]$ is set to 1 if an only if $x[m-i..m-1-i+k] = y[j+m-k..j+m-1]$. At the beginning of each attempt, d is set to 1^{m-1}. When reading text character $y[j+m-k]$, for $1 \le k \le m$, d takes the following value $\text{Shift}(d \text{ And } B[y[j+m-k]])$.

There is a match if and only if, after iteration m, it holds $d[m-1] = 1$. If during an attempt at left position j, the m characters of the window are scanned and d never takes the value 0 then an occurrence of x occurs in y at position j.

Whenever $d[m-1] = 1$, the algorithm has matched a prefix of the pattern in the current window position j. The longest prefix matched gives the shift to the next position.

The preprocessing phase of the Backward Nondeterministic Dawg Matching algorithm consists in computing the array B. The Backward Nondeterministic Dawg Matching algorithm has a quadratic worst case time complexity but it is optimal in average.

3 The C code

```
void BNDM(String x, int m, String y, int n) {
   unsigned int s, d, B[ASIZE];
   int i, j, last;
   if (m > WORD_SIZE) error("BNDM");
   /* Preprocessing */
   memset(B, 0, ASIZE*sizeof(int));
   s = 1;
   for (i = m - 1; i >= 0; i--) {
      B[x[i]] |= s;
      s <<= 1;
   }
   s >>= 1;
   /* Searching */
   j = 0;
   while (j <= n-m) {
      i = m-1;
      last = m;
      d = ~0;
      while (i >= 0 && d != 0) {
         d &= B[y[j+i]];
         i--;
         if (d != 0)
            if (i < 0)
               OUTPUT(j);
            else
               if ((d & s) != 0)
                  last = i+1;
         d <<= 1;
      }
      j += last;
   }
}
```

4 The example

c	B[c]
A	00101010
C	01000000
G	10010101
T	00000000

32. BACKWARD NONDETERMINISTIC DAWG MATCHING ALGORITHM 175

Searching phase

First attempt:

y | G C A T C G C A | G A G A G T A T A C A G T A C G |

$d = 11111111 \, And \, B[A] = 00101010$

y | G C A T C G C A | G A G A G T A T A C A G T A C G |

$d = 01010100 \, And \, B[C] = 01000000$

y | G C A T C G C A | G A G A G T A T A C A G T A C G |

$d = 10000000 \, And \, B[G] = 10000000$

y | G C A T C G C A | G A G A G T A T A C A G T A C G |

$d = 00000000 \, And \, B[C] = 00000000$

Shift by 5 (8 − 3)

Second attempt:

y | G C A T C G C A G A G A G T A T A C A G T A C G |

$d = 11111111 \, And \, B[G] = 10010101$

y | G C A T C G C A G A G A G T A T A C A G T A C G |

$d = 00101010 \, And \, B[A] = 00101010$

y | G C A T C G C A G A G A G T A T A C A G T A C G |

$d = 01010100 \, And \, B[G] = 00010100$

y `G C A T C `|`G C A G A G A G`|`T A T A C A G T A C G`

$d = 00101000$ And $B[\text{A}] = 00101000$

y `G C A T C `|`G C A G A G A G`|`T A T A C A G T A C G`

$d = 01010000$ And $B[\text{G}] = 00010000$

y `G C A T C `|`G C A G A G A G`|`T A T A C A G T A C G`

$d = 00100000$ And $B[\text{A}] = 00100000$

y `G C A T C `|`G C A G A G A G`|`T A T A C A G T A C G`

$d = 01000000$ And $B[\text{C}] = 01000000$

y `G C A T C `|`G C A G A G A G`|`T A T A C A G T A C G`

$d = 10000000$ And $B[\text{G}] = 10000000$

Shift by 7 (8-1)

Third attempt:

y `G C A T C G C A G A G A `|`G T A T A C A G`|`T A C G`

$d = 11111111$ And $B[\text{G}] = 10010101$

y `G C A T C G C A G A G A `|`G T A T A C A G`|`T A C G`

$d = 10010101$ And $B[\text{A}] = 00101010$

y `G C A T C G C A G A G A `|`G T A T A C A G`|`T A C G`

$d = 01010100$ And $B[\text{C}] = 00010100$

y | G C A T C G C A G A G A | G T A T A C A G | T A C G |

$d = 00101000$ And $B[\text{A}] = 00000000$

Shift by 7 (8-1)

The Backward Nondeterministic Dawg Matching algorithm performs 16 text character inspections on the example.

5 References

- NAVARRO, G., RAFFINOT, M., 1998, A bit-parallel approach to suffix automata: fast extended string matching, in *Proceedings of the 9th Annual Symposium on Combinatorial Pattern Matching*, M. Farach-Colton ed., Piscataway, New Jersey, Lecture Notes in Computer Science 1448, pp 14–31, Springer-Verlag, Berlin.

- NAVARRO, G., RAFFINOT, M., 2002, *Flexible Pattern Matching in Strings Practical on-line search algorithms for texts and biological sequences*, Cambridge University Press.

CHAPTER 33

GALIL-SEIFERAS ALGORITHM

1 Main features

- constant extra space complexity;
- preprocessing phase in $O(m)$ time and constant space complexity;
- searching phase in $O(n)$ time complexity;
- performs $5n$ text character comparisons in the worst case.

2 Description

Throughout this chapter we will use a constant k. Galil and Seiferas suggest that practically this constant could be equal to 4.

Let us define the function *reach* for $0 \leq i < m$ as follows:

$$reach(i) = i + \max\{i' \leq m - i \mid x[0..i'] = x[i+1..i'+i+1]\}\,.$$

Then a prefix $x[0..p]$ of x is a **prefix period** if it is basic and $reach(p) \geq k \times p$.

The preprocessing phase of the Galil-Seiferas algorithm consists in finding a decomposition uv of x such that v has at most one prefix period and $|u| = O(per(v))$. Such a decomposition is called a **perfect factorization**.

Then the searching phase consists in scanning the text y for every occurrences of v and when v occurs to check naively if u occurs just before in y.

In the implementation below the aim of the preprocessing phase (functions newP1, newP2 and parse) is to find a perfect factorization uv of x where $u = x[0..s-1]$ and $v = x[s..m-1]$. Function newP1 finds the shortest prefix period of $x[s..m-1]$. Function newP2 finds the second shortest prefix period of $x[s..m-1]$ and function parse increments s.

Before calling function search we have:

- $x[s..m-1]$ has at most one prefix period;

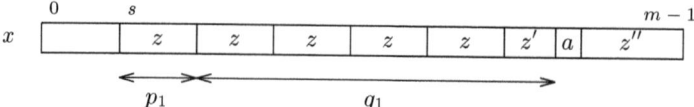

Figure 33.1. A perfect factorization of x.

- if $x[s \mathinner{.\,.} m-1]$ does have a prefix period, then its length is p_1;
- $x[s \mathinner{.\,.} s+p_1+q_1-1]$ has shortest period of length p_1;
- $x[s \mathinner{.\,.} s+p_1+q_1]$ does not have period of length p_1.

The pattern x is of the form $x[0 \mathinner{.\,.} s-1]x[s \mathinner{.\,.} m-1]$ where $x[s \mathinner{.\,.} m-1]$ is of the form $z^\ell z' a z''$ with z basic, $|z| = p_1$, z' prefix of z, $z'a$ not a prefix of z and $|z^\ell z'| = p_1 + q_1$ (see figure 33.1).

It means that when searching for $x[s \mathinner{.\,.} m-1]$ in y:

- if $x[s \mathinner{.\,.} s+p_1+q_1-1]$ has been matched a shift of length p_1 can be performed and the comparisons are resumed with $x[s+q_1]$;
- otherwise if a mismatch occurs with $x[s+q]$ with $q \neq p_1 + q_1$ then a shift of length $q/k + 1$ can be performed and the comparisons are resumed with $x[0]$.

This gives an overall linear number of text character comparisons.

The preprocessing phase of the Galil-Seiferas algorithm is in $O(m)$ time and constant space complexity. The searching phase is in $O(n)$ time complexity. At most $5n$ text character comparisons can be done during this phase.

3 The C code

All the variables are global.

```
String x, y;
int k, m, n, p, p1, p2, q, q1, q2, s;


void search() {
   while (p <= n - m) {
      while (p + s + q < n && x[s + q] == y[p + s + q])
         ++q;
      if (q == m - s && memcmp(x, y + p, s + 1) == 0)
         OUTPUT(p);
```

```
         if (q == p1 + q1) {
            p += p1;
            q -= p1;
         }
         else {
            p += (q/k + 1);
            q = 0;
         }
      }
   }
}


void parse() {
   while (1) {
      while (x[s + q1] == x[s + p1 + q1])
         ++q1;
      while (p1 + q1 >= k*p1) {
         s += p1;
         q1 -= p1;
      }
      p1 += (q1/k + 1);
      q1 = 0;
      if (p1 >= p2)
         break;
   }
   newP1();
}


void newP2() {
   while (x[s + q2] == x[s + p2 + q2] && p2 + q2 < k*p2)
      ++q2;
   if (p2 + q2 == k*p2)
      parse();
   else
      if (s + p2 + q2 == m)
         search();
      else {
         if (q2 == p1 + q1) {
            p2 += p1;
            q2 -= p1;
```

```
         }
         else {
            p2 += (q2/k + 1); q2 = 0;
         }
         newP2();
      }
}


void newP1() {
   while (x[s + q1] == x[s + p1 + q1])
      ++q1;
   if (p1 + q1 >= k*p1) {
      p2 = q1;
      q2 = 0;
      newP2();
   }
   else {
      if (s + p1 + q1 == m)
         search();
      else {
         p1 += (q1/k + 1);
         q1 = 0;
         newP1();
      }
   }
}


void GS(String argX, int argM, String argY, int argN) {
   x = argX;
   m = argM;
   y = argY;
   n = argN;
   k = 4;
   p = q = s = q1 = p2 = q2 = 0;
   p1 = 1;
   newP1();
}
```

4 The example

$p = 0, q = 0, s = 0, p_1 = 7, q_1 = 1$

Searching phase

First attempt:

```
y  G C A T C G C A G A G A G T A T A C A G T A C G
   1 2 3 4
x  G C A G A G A G
```
Shift by 1

Second attempt:

```
y  G C A T C G C A G A G A G T A T A C A G T A C G
     1
x    G C A G A G A G
```
Shift by 1

Third attempt:

```
y  G C A T C G C A G A G A G T A T A C A G T A C G
       1
x      G C A G A G A G
```
Shift by 1

Fourth attempt:

```
y  G C A T C G C A G A G A G T A T A C A G T A C G
         1
x        G C A G A G A G
```
Shift by 1

Fifth attempt:

```
y  G C A T C G C A G A G A G T A T A C A G T A C G
           1
x          G C A G A G A G
```
Shift by 1

Sixth attempt:

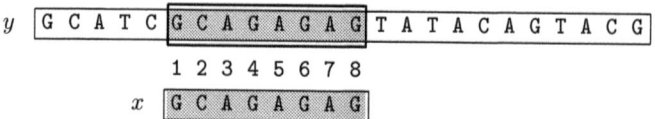

Shift by 7 (p_1)

Seventh attempt:

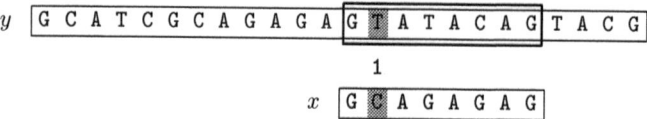

Shift by 1

Eighth attempt:

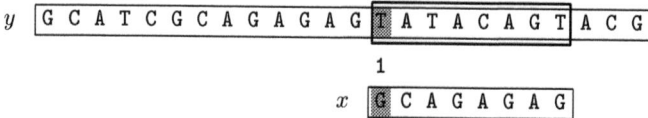

Shift by 1

Ninth attempt:

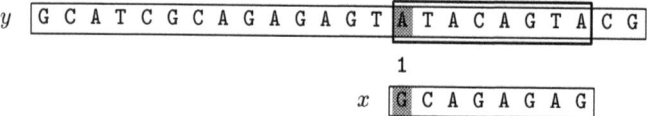

Shift by 1

Tenth attempt:

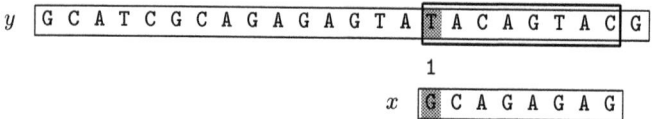

Shift by 1

Eleventh attempt:

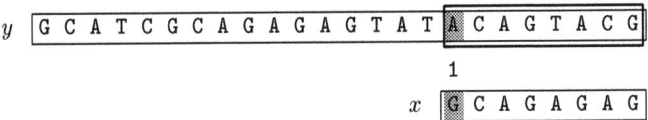

Shift by 1

The Galil-Seiferas algorithm performs 21 text character comparisons on the example.

5 References

- CROCHEMORE, M., RYTTER, W., 1994, *Text Algorithms*, Oxford University Press.

- CROCHEMORE, M., RYTTER, W., 2002, *Jewels of Stringology*, World Scientific Press.

- GALIL, Z., SEIFERAS, J., 1983, Time-space optimal string matching, *Journal of Computer and System Science* **26**(3):280–294.

CHAPTER 34

TWO WAY ALGORITHM

1 Main features

- requires an ordered alphabet;
- preprocessing phase in $O(m)$ time and constant space complexity;
- searching phase in $O(n)$ time;
- performs $2n - m$ text character comparisons in the worst case.

2 Description

The pattern x is factorized into two parts x_ℓ and x_r such that $x = x_\ell x_r$. Then the searching phase of the Two Way algorithm consists in comparing the characters of x_r from left to right and then, if no mismatch occurs during that first stage, in comparing the characters of x_ℓ from right to left in a second stage.

The preprocessing phase of the algorithm consists then in choosing a good **factorization** $x_\ell x_r$. Let (u, v) be a factorization of x. A **repetition** in (u, v) is a word w such that the two following properties hold:

(i) w is a suffix of u or u is a suffix of w;

(ii) w is a prefix of v of v is a prefix of w.

In other words w occurs at both sides of the cut between u and v with a possible overflow on either side. The length of a repetition in (u, v) is called a **local period** and the length of the smallest repetition in (u, v) is called **the local period** and is denoted by $r(u, v)$.

Each factorization (u, v) of x has at least one repetition. It can be easily seen that

$$1 \leq r(u, v) \leq |x|.$$

A factorization (u, v) of x such that $r(u, v) = \mathrm{per}(x)$ is called a **critical factorization** of x.

If (u,v) is a critical factorization of x then at the position $|u|$ in x the global and the local periods are the same. The Two Way algorithm chooses the critical factorization (x_ℓ, x_r) such that $|x_\ell| < per(x)$ and $|x_\ell|$ is minimal.

To compute the critical factorization (x_ℓ, x_r) of x we first compute the maximal suffix z of x for the order $\leq$ and the maximal suffix $\tilde{z}$ for the reverse order $\tilde{\leq}$. Then (x_ℓ, x_r) is chosen such that $|x_\ell| = \max\{|z|, |\tilde{z}|\}$.

The preprocessing phase can be done in $O(m)$ time and constant space complexity.

The searching phase of the Two Way algorithm consists in first comparing the character of x_r from left to right, then the character of x_ℓ from right to left.

When a mismatch occurs when scanning the k-th character of x_r, then a shift of length k is performed.

When a mismatch occurs when scanning x_ℓ or when an occurrence of the pattern is found, then a shift of length $per(x)$ is performed.

Such a scheme leads to a quadratic worst case algorithm, this can be avoided by a prefix memorization: when a shift of length $per(x)$ is performed the length of the matching prefix of the pattern at the beginning of the window (namely $m - per(x)$) after the shift is memorized to avoid to scan it again during the next attempt.

The searching phase of the Two Way algorithm can be done in $O(n)$ time complexity. The algorithm performs $2n - m$ text character comparisons in the worst case. Breslauer designed a variation of the Two Way algorithm which performs less than $2n - m$ comparisons in the worst case using constant space.

3 The C code

```
/* Computing of the maximal suffix for <= */
int maxSuf(String x, int m, int *p) {
   int ms, j, k;
   Character a, b;

   ms = -1;
   j = 0;
   k = *p = 1;
   while (j + k < m) {
      a = x[j + k];
      b = x[ms + k];
      if (a < b) {
         j += k;
         k = 1;
```

```
               *p = j - ms;
         }
         else
            if (a == b)
               if (k != *p)
                  ++k;
               else {
                  j += *p;
                  k = 1;
               }
            else { /* a > b */
               ms = j;
               j = ms + 1;
               k = *p = 1;
            }
      }
      return(ms);
}

/* Computing of the maximal suffix for >= */
int maxSufTilde(String x, int m, int *p) {
   int ms, j, k;
   Character a, b;

   ms = -1;
   j = 0;
   k = *p = 1;
   while (j + k < m) {
      a = x[j + k];
      b = x[ms + k];
      if (a > b) {
         j += k;
         k = 1;
         *p = j - ms;
      }
      else
         if (a == b)
            if (k != *p)
               ++k;
            else {
               j += *p;
```

```
               k = 1;
         }
         else { /* a < b */
            ms = j;
            j = ms + 1;
            k = *p = 1;
         }
      }
      return(ms);
}


/* Two Way string matching algorithm. */
void TW(String x, int m, String y, int n) {
   int i, j, ell, memory, p, per, q;

   /* Preprocessing */
   i = maxSuf(x, m, &p);
   j = maxSufTilde(x, m, &q);
   if (i > j) {
      ell = i;
      per = p;
   }
   else {
      ell = j;
      per = q;
   }

   /* Searching */
   if (memcmp(x, x + per, ell + 1) == 0) {
      j = 0;
      memory = -1;
      while (j <= n - m) {
         i = MAX(ell, memory) + 1;
         while (i < m && x[i] == y[i + j])
            ++i;
         if (i >= m) {
            i = ell;
            while (i > memory && x[i] == y[i + j])
               --i;
            if (i <= memory)
```

```
            OUTPUT(j);
         j += per;
         memory = m - per - 1;
      }
      else {
         j += (i - ell);
         memory = -1;
      }
   }
}
else {
   per = MAX(ell + 1, m - ell - 1) + 1;
   j = 0;
   while (j <= n - m) {
      i = ell + 1;
      while (i < m && x[i] == y[i + j])
         ++i;
      if (i >= m) {
         i = ell;
         while (i >= 0 && x[i] == y[i + j])
            --i;
         if (i < 0)
            OUTPUT(j);
         j += per;
      }
      else
         j += (i - ell);
   }
}
}
```

4 The example

x G C A G A G A G
local period   1 3 7 7 2 2 2 1

$x_\ell = $ GC, $x_r = $ AGAGAG

Searching phase

First attempt:

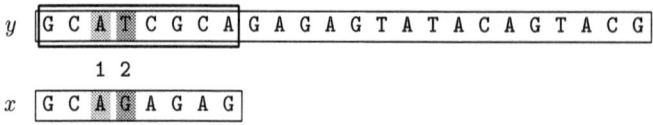

Shift by 2

Second attempt:

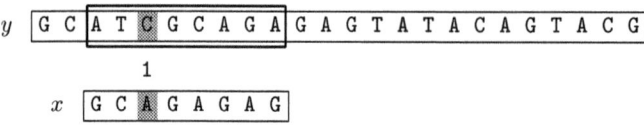

Shift by 1

Third attempt:

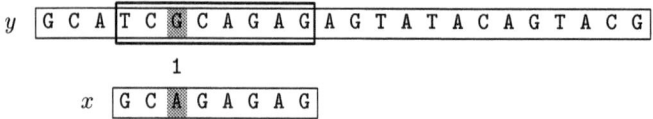

Shift by 1

Fourth attempt:

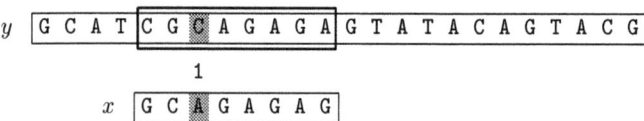

Shift by 1

Fifth attempt:

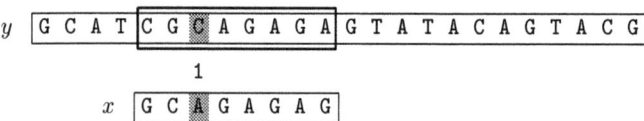

Shift by 7

Sixth attempt:

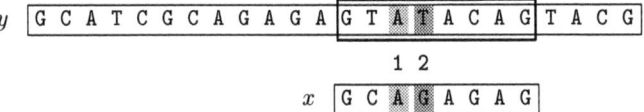

Shift by 2

Seventh attempt:

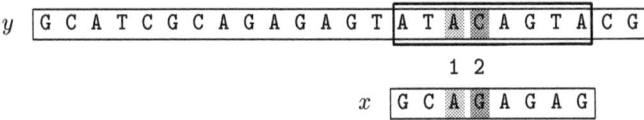

Shift by 2

Eighth attempt:

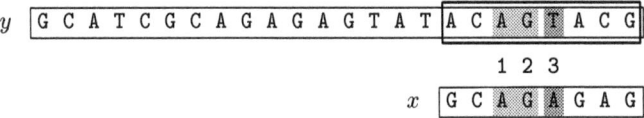

Shift by 3

The Two Way algorithm performs 20 text character comparisons on the example.

5 References

- BRESLAUER, D., 1996, Saving comparisons in the Crochemore-Perrin string matching algorithm, *Theoretical Computer Science* **158**(1–2):177–192.

- CROCHEMORE, M., 1997, Off-line serial exact string searching, in *Pattern Matching Algorithms*, A. Apostolico and Z. Galil eds., Chapter 1, pp 1–53, Oxford University Press.

- CROCHEMORE, M., PERRIN, D., 1991, Two-way string-matching, *Journal of the ACM* **38**(3):651–675.

- CROCHEMORE, M., RYTTER, W., 1994, *Text Algorithms*, Oxford University Press.

- CROCHEMORE, M., RYTTER, W., 2002, *Jewels of Stringology*, World Scientific Press.

CHAPTER 35

STRING MATCHING ON ORDERED ALPHABETS

1 Main features

- no preprocessing phase;
- requires an ordered alphabet;
- constant extra space complexity;
- searching phase in $O(n)$ time;
- performs $6n+5$ text character comparisons in the worst case.

2 Description

Consider an attempt where the window is positioned on the text factor $y[j \mathinner{.\,.} j+m-1]$, when a prefix u of x has been matched and a mismatch occurs between characters a in x and b in y (see figure 35.1), the algorithm tries to compute the period of ub, if it does not succeed in finding the exact period it computes an approximation of it.

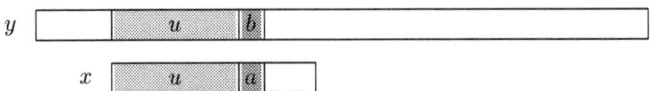

Figure 35.1. Typical attempt during the String Matching on Ordered Alphabets algorithm.

Let us define $tw^e w'$ the **Maximal-Suffix decomposition** (MS decomposition for short) of the word x such that:

- $v = w^e w'$ is the maximal suffix of x according to the alphabetical ordering;
- w is basic;

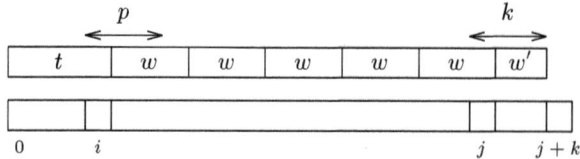

Figure 35.2. Function nextMaximalSuffix: meaning of the variables i, j, k and p.

- $e \geq 1$;
- w' is a proper prefix of w.

Then we have $|t| < per(x)$.

If $tw^e w'$ is the MS decomposition of a nonempty word x then the four following properties hold:

- if t is a suffix of w then $per(x) = per(v)$;
- $per(x) > |t|$;
- if $|t| \geq |w|$ then $per(x) > |v| = |x| - |t|$;
- if t is not a suffix of w and $|t| < |w|$ then $per(x) > \min(|v|, |tw^e|)$.

If t is a suffix of w then $per(x) = per(v) = |w|$.
Otherwise $per(x) > \max(|t|, \min(|v|, |tw^e|)) \geq |x|/2$.

If $tw^e w'$ is the MS decomposition of a nonempty word x, $per(x) = |w|$ and $e > 1$ then $tw^{e-1} w'$ is the MS decomposition of $x' = tw^{e-1} w'$.

The algorithm computes the maximal suffix of the matched prefix of the pattern appended with the mismatched character of the text after each attempt. It avoids to compute it from scratch after a shift of length $per(w)$ has been performed.

The algorithm String Matching on Ordered Alphabets needs no preprocessing phase.

The searching phase can be done in $O(n)$ time complexity using a constant extra space. The algorithm performs no more than $6n+5$ text character comparisons.

3 The C code

Figure 35.2 gives the meaning of the four variables i, j, k and p in the function nextMaximalSuffix: $i = |t| - 1$, $j = |tw^e| - 1$, $k = |w'| + 1$ and $p = |w|$.

35. STRING MATCHING ON ORDERED ALPHABETS

```c
/* Compute the next maximal suffix. */
void nextMaximalSuffix(String x, int m,
                      int *i, int *j, int *k, int *p) {
   Character a, b;

   while (*j + *k < m) {
      a = x[*i + *k];
      b = x[*j + *k];
      if (a == b)
         if (*k == *p) {
            (*j) += *p;
            *k = 1;
         }
         else
            ++(*k);
      else
         if (a > b) {
            (*j) += *k;
            *k = 1;
            *p = *j - *i;
         }
         else {
            *i = *j;
            ++(*j);
            *k = *p = 1;
         }
   }
}


/* String matching on ordered alphabets algorithm. */
void SMOA(String x, int m, String y, int n) {
   int i, ip, j, jp, k, p;

   /* Searching */
   ip = -1;
   i = j = jp = 0;
   k = p = 1;
   while (j <= n - m) {
      while (i + j < n && i < m && x[i] == y[i + j])
         ++i;
```

```
         if (i == 0) {
            ++j;
            ip = -1;
            jp = 0;
            k = p = 1;
         }
         else {
            if (i >= m)
               OUTPUT(j);
            nextMaximalSuffix(y + j, i+1, &ip, &jp, &k, &p);
            if (ip < 0 ||
                (ip < p &&
                 memcmp(y + j, y + j + p, ip + 1) == 0)) {
               j += p;
               i -= p;
               if (i < 0)
                  i = 0;
               if (jp - ip > p)
                  jp -= p;
               else {
                  ip = -1;
                  jp = 0;
                  k = p = 1;
               }
            }
            else {
               j += (MAX(ip + 1,
                         MIN(i - ip - 1, jp + 1)) + 1);
               i = jp = 0;
               ip = -1;
               k = p = 1;
            }
         }
      }
   }
}
```

4 The example
Searching phase

First attempt:

```
y  G C A T C G C A G A G A G T A T A C A G T A C G
   1 2 3 4
x  G C A G A G A G
```

After a call of nextMaximalSuffix: $ip = 2, jp = 3, k = 1, p = 1$. It performs 6 text character comparisons.
Shift by 4

Second attempt:

```
y  G C A T C G C A G A G A G T A T A C A G T A C G
           1
x        G C A G A G A G
```

Shift by 1

Third attempt:

```
y  G C A T C G C A G A G A G T A T A C A G T A C G
           1 2 3 4 5 6 7 8
x          G C A G A G A G
```

After a call of nextMaximalSuffix: $ip = 7, jp = 8, k = 1, p = 1$. It performs 15 text character comparisons.
Shift by 9

Fourth attempt:

```
y  G C A T C G C A G A G A G T A T A C A G T A C G
                              1
x                          G C A G A G A G
```

Shift by 1

Fifth attempt:

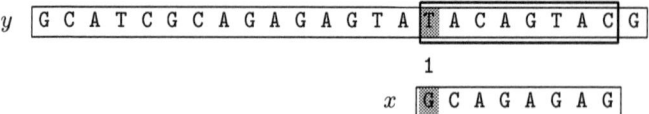

Shift by 1

Sixth attempt:

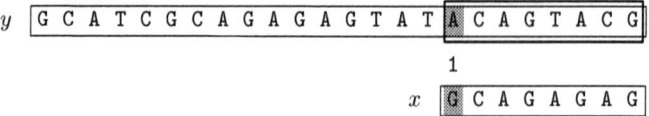

Shift by 1

The string matching on ordered alphabets algorithm performs 37 text character comparisons on the example.

5 References

- CROCHEMORE, M., 1992, String-matching on ordered alphabets, *Theoretical Computer Science* **92**(1):33–47.

- CROCHEMORE, M., RYTTER, W., 1994, *Text Algorithms*, Oxford University Press.

CHAPTER 36

OPTIMAL MISMATCH ALGORITHM

1 Main features
- variant of the Quick Search algorithm;
- requires the frequencies of the characters;
- preprocessing phase in $O(m^2+\sigma)$ time and $O(m+\sigma)$ space complexity;
- searching phase in $O(m \times n)$ time complexity.

2 Description

Sunday designed an algorithm where the pattern characters are scanned from the least frequent one to the most frequent one. Doing so one may hope to have a mismatch most of the times and thus to scan the whole text very quickly. One needs to know the frequencies of each of the character of the alphabet.

The preprocessing phase of the Optimal Mismatch algorithm consists in sorting the pattern characters in decreasing order of their frequencies and then in building the Quick Search bad-character shift function (see chapter 22) and a good-suffix shift function adapted to the scanning order of the pattern characters. It can be done in $O(m^2 + \sigma)$ time and $O(m + \sigma)$ space complexity.

The searching phase of the Optimal Mismatch algorithm has a $O(m \times n)$ time complexity.

3 The C code

The function preQsBc is given chapter 22.

```
typedef struct patternScanOrder {
   int loc;
   Character c;
} pattern;
```

```
int freq[ASIZE];

/* Construct an ordered pattern from a string. */
void orderPattern(String x, int m, int (*pcmp)(),
                  pattern *pat) {
   int i;

   for (i = 0; i <= m; ++i) {
      pat[i].loc = i;
      pat[i].c = x[i];
   }
   qsort(pat, m, sizeof(pattern), pcmp);
}

/* Optimal Mismatch pattern comparison function. */
int optimalPcmp(pattern *pat1, pattern *pat2) {
   float fx;

   fx = freq[pat1->c] - freq[pat2->c];
   return(fx ? (fx > 0 ? 1 : -1) :
               (pat2->loc - pat1->loc));
}

/* Find the next leftward matching shift for the first ploc
 pattern elements after a current shift or lshift. */
int matchShift(String x, int m, int ploc,
               int lshift, pattern *pat) {
   int i, j;

   for (; lshift < m; ++lshift) {
      i = ploc;
      while (--i >= 0) {
         if ((j = (pat[i].loc - lshift)) < 0)
            continue;
         if (pat[i].c != x[j])
            break;
      }
      if (i < 0)
         break;
   }
   return(lshift);
```

```
}

/* Constructs the good-suffix shift table
   from an ordered string. */
void preAdaptedGs(String x, int m, int adaptedGs[],
                  pattern *pat) {
   int lshift, i, ploc;

   adaptedGs[0] = lshift = 1;
   for (ploc = 1; ploc <= m; ++ploc) {
      lshift = matchShift(x, m, ploc, lshift, pat);
      adaptedGs[ploc] = lshift;
   }
   for (ploc = 0; ploc <= m; ++ploc) {
      lshift = adaptedGs[ploc];
      while (lshift < m) {
         i = pat[ploc].loc - lshift;
         if (i < 0 || pat[ploc].c != x[i])
            break;
         ++lshift;
         lshift = matchShift(x, m, ploc, lshift, pat);
      }
      adaptedGs[ploc] = lshift;
   }
}

/* Optimal Mismatch string matching algorithm. */
void OM(String x, int m, String y, int n) {
   int i, j, adaptedGs[XSIZE], qsBc[ASIZE];
   pattern pat[XSIZE];

   /* Preprocessing */
   orderPattern(x, m, optimalPcmp, pat);
   preQsBc(x, m, qsBc);
   preAdaptedGs(x, m, adaptedGs, pat);

   /* Searching */
   j = 0;
   while (j <= n - m) {
      i = 0;
      while (i < m && pat[i].c == y[j + pat[i].loc])
```

```
        ++i;
    if (i >= m)
        OUTPUT(j);
    j += MAX(adaptedGs[i],qsBc[y[j + m]]);
  }
}
```

4 The example

c	A	C	G	T
freq[c]	8	5	7	4
qsBc[c]	2	7	1	9

i	0	1	2	3	4	5	6	7
x[i]	G	C	A	G	A	G	A	G
pat[i].loc	1	7	5	3	0	6	4	2
pat[i].c	C	G	G	G	G	A	A	A

i	0	1	2	3	4	5	6	7	8
adaptedGs[i]	1	3	4	2	7	7	7	7	7

Searching phase

First attempt:

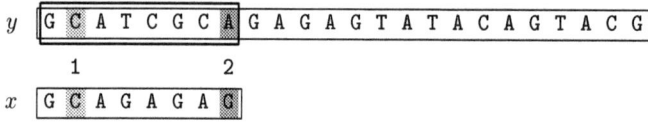

Shift by 3 (adaptedGs[1])

Second attempt:

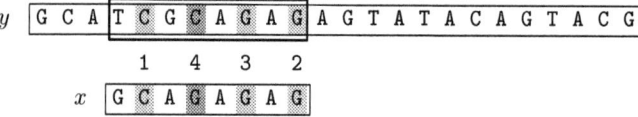

Shift by 2 (qsBc[A] = adaptedGs[3])

Fourth attempt:

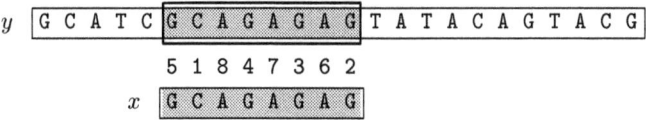

Shift by 9 ($qsBc[\text{T}]$)

Fifth attempt:

```
y  G C A T C G C A G A G A G T A T A C A G T A C G
                                   1
                             x  G C A G A G A G
```

Shift by 7 ($qsBc[\text{C}]$)

The Optimal Mismatch algorithm performs 15 text character comparisons and it inspects 4 more text characters in order to compute the shifts, on the example.

5 References

- SMYTH, W. F., 2003, *Computing Patterns in Strings*, Pearson Addison Wesley.

- SUNDAY, D.M., 1990, A very fast substring search algorithm, *Communications of the ACM* **33**(8):132–142.

CHAPTER 37

MAXIMAL SHIFT ALGORITHM

1 Main features

- variant of the Quick Search algorithm;
- quadratic worst case time complexity;
- preprocessing phase in $O(m^2+\sigma)$ time and $O(m+\sigma)$ space complexity;
- searching phase in $O(m \times n)$ time complexity.

2 Description

Sunday designed an algorithm where the pattern characters are scanned from the one which will lead to a larger shift to the one which will lead to a shorter shift. Doing so one may hope to maximize the lengths of the shifts.

The preprocessing phase of the Maximal Shift algorithm consists in sorting the pattern characters in decreasing order of their shift and then in building the Quick Search bad-character shift function (see chapter 22) and a good-suffix shift function adapted to the scanning order of the pattern characters. It can be done in $O(m^2 + \sigma)$ time and $O(m+\sigma)$ space complexity.

The searching phase of the Maximal Shift algorithm has a quadratic worst case time complexity.

3 The C code

The function preQsBc is given chapter 22. The functions orderPattern, matchShift and preAdaptedGs are given chapter 36.

```
typedef struct patternScanOrder {
    int loc;
    Character c;
} pattern;

int minShift[XSIZE];
```

```
/* Computation of the MinShift table values. */
void computeMinShift(String x, int m) {
   int i, j;

   for (i = 0; i < m; ++i) {
      for (j = i - 1; j >= 0; --j)
         if (x[i] == x[j]) break;
      minShift[i] = i - j;
   }
}

/* Maximal Shift pattern comparison function. */
int maxShiftPcmp(pattern *pat1, pattern *pat2) {
   int dsh;

   dsh = minShift[pat2->loc] - minShift[pat1->loc];
   return(dsh ? dsh : (pat2->loc - pat1->loc));
}

/* Maximal Shift string matching algorithm. */
void MS(String x, int m, String y, int n) {
   int i, j, qsBc[ASIZE], adaptedGs[XSIZE];
   pattern pat[XSIZE];

   /* Preprocessing */
   computeMinShift(x ,m);
   orderPattern(x, m, maxShiftPcmp, pat);
   preQsBc(x, m, qsBc);
   preAdaptedGs(x, m, adaptedGs, pat);
   /* Searching */
   j = 0;
   while (j <= n - m) {
      i = 0;
      while (i < m && pat[i].c == y[j + pat[i].loc])
         ++i;
      if (i >= m)
         OUTPUT(j);
      j += MAX(adaptedGs[i], qsBc[y[j + m]]);
   }
}
```

4 The example

i	0	1	2	3	4	5	6	7
$x[i]$	G	C	A	G	A	G	A	G
$minShift[i]$	1	2	3	3	2	2	2	2
$pat[i].loc$	3	2	7	6	5	4	1	0
$pat[i].c$	G	A	G	A	G	A	C	G

c	A	C	G	T
$qsBc[c]$	2	7	1	9

i	0	1	2	3	4	5	6	7	8
$adaptedGs[i]$	1	3	3	7	4	7	7	7	7

Searching phase

First attempt:

```
y  G C A T C G C A G A G A G T A T A C A G T A C G
            1
x  G C A G A G A G
```

Shift by 1 ($qsBc[\text{G}] = adaptedGs[0]$)

Second attempt:

```
y  G C A T C G C A G A G A G T A T A C A G T A C G
          1
   x  G C A G A G A G
```

Shift by 2 ($qsBc[\text{A}]$)

Third attempt:

```
y  G C A T C G C A G A G A G T A T A C A G T A C G
              1
     x  G C A G A G A G
```

Shift by 2 ($qsBc[\text{A}]$)

Fourth attempt:

```
y  G C A T C G C A G A G A G T A T A C A G T A C G
           8 7 2 1 6 5 4 3
        x  G C A G A G A G
```

Shift by 9 ($qsBc[\text{T}]$)

Fifth attempt:

```
y  G C A T C G C A G A G A G T A T A C A G T A C G
                                         1
                          x  G C A G A G A G
```

Shift by 7 ($qsBc[\text{C}]$)

The Maximal Shift algorithm performs 12 text character comparisons it inspects 4 more text characters in order to compute the shifts, on the example.

5 References

- SMYTH, W. F., 2003, *Computing Patterns in Strings*, Pearson Addison Wesley.

- SUNDAY, D.M., 1990, A very fast substring search algorithm, *Communications of the ACM* **33**(8):132–142.

CHAPTER 38

SKIP SEARCH ALGORITHM

1 Main features

- uses buckets of positions for each character of the alphabet;
- preprocessing phase in $O(m + \sigma)$ time and space complexity;
- searching phase in $O(m \times n)$ time complexity;
- $O(n)$ expected text character comparisons.

2 Description

For each character of the alphabet, a bucket collects all the positions of that character in x. When a character occurs k times in the pattern, there are k corresponding positions in the bucket of the character. When the word is much shorter than the alphabet, many buckets are empty.

The preprocessing phase of the Skip Search algorithm consists in computing the buckets for all the characters of the alphabet: for $c \in \Sigma$

$$z[c] = \{i \mid 0 \le i \le m - 1 \text{ and } x[i] = c\}.$$

The space and time complexity of this preprocessing phase is $O(m + \sigma)$.

The main loop of the search phase consists in examining every m-th text character, $y[j]$ (so there will be n/m main iterations). For $y[j]$, it uses each position in the bucket $z[y[j]]$ to obtain a possible starting position p of x in y. It performs a comparison of x with y beginning at position p, character by character, until there is a mismatch, or until all match.

The Skip Search algorithm has a quadratic worst case time complexity but the expected number of text character inspections is $O(n)$.

3 The C code

The description of a linked list List can be found section 5.

```
void SKIP(String x, int m, String y, int n) {
   int i, j;
   List ptr, z[ASIZE];

   /* Preprocessing */
   memset(z, NULL, ASIZE*sizeof(List));
   for (i = 0; i < m; ++i) {
      ptr = (List)malloc(sizeof(struct _cell));
      if (ptr == NULL)
         error("SKIP");
      ptr->element = i;
      ptr->next = z[x[i]];
      z[x[i]] = ptr;
   }

   /* Searching */
   for (j = m - 1; j < n; j += m)
      for (ptr = z[y[j]]; ptr != NULL; ptr = ptr->next)
         if (memcmp(x, y + j - ptr->element, m) == 0) {
            if (j - ptr->element <= n - m)
               OUTPUT(j - ptr->element);
         }
         else break;
}
```

In practice the test `j - ptr->element <= n - m` can be omitted and the algorithm becomes :

```
void SKIP(String x, int m, String y, int n) {
   int i, j;
   List ptr, z[ASIZE];

   /* Preprocessing */
   memset(z, NULL, ASIZE*sizeof(List));
   for (i = 0; i < m; ++i) {
      ptr = (List)malloc(sizeof(struct _cell));
      if (ptr == NULL)
         error("SKIP");
      ptr->element = i;
      ptr->next = z[x[i]];
      z[x[i]] = ptr;
   }
```

```
    /* Searching */
    for (j = m - 1; j < n; j += m)
       for (ptr = z[y[j]]; ptr != NULL; ptr = ptr->next)
          if (memcmp(x, y + j - ptr->element, m) == 0)
             OUTPUT(j - ptr->element);
}
```

4 The example

c	$z[c]$
A	$(6, 4, 2)$
C	(1)
G	$(7, 5, 3, 0)$
T	$\emptyset$

Searching phase

First attempt:

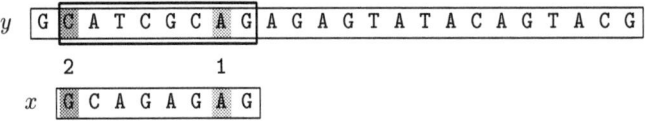

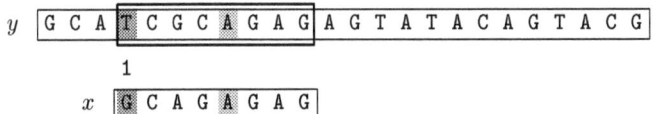

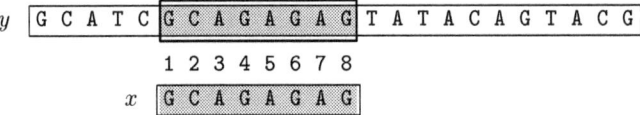

Shift by 8

Second attempt:

| y | G C A T C G A G A G A G T A T A C A G T A C G |

 1

Shift by 8

Third attempt:

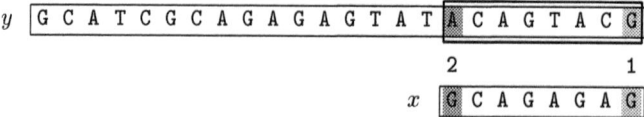

The Skip Search algorithm performs 14 text character inspections on the example.

5 References

- CHARRAS, C., LECROQ, T., PEHOUSHEK, J.D., 1998, A very fast string matching algorithm for small alphabets and long patterns, in *Proceedings of the 9th Annual Symposium on Combinatorial Pattern Matching*, M. Farach-Colton ed., Piscataway, New Jersey, Lecture Notes in Computer Science 1448, pp 55–64, Springer-Verlag, Berlin.

- LECROQ, T., 2000, *Quelques aspects de l'algorithmique du texte*, Habilitation thesis, Université de Rouen, France.

CHAPTER 39

KMPSKIP SEARCH ALGORITHM

1 Main features

- improvement of the Skip Search algorithm;
- uses buckets of positions for each character of the alphabet;
- preprocessing phase in $O(m + \sigma)$ time and space complexity;
- searching phase in $O(n)$ time complexity.

2 Description

It is possible to make the Skip Search algorithm (see chapter 38) linear using the two shift tables of Morris-Pratt (see chapter 6) and Knuth-Morris-Pratt (see chapter 7).

For $1 \leq i \leq m$, $mpNext[i]$ is equal to the length of the longest border of $x[0..i-1]$ and $mpNext[0] = -1$.

For $1 \leq i < m$, $kmpNext[i]$ is equal to length of the longest border of $x[0..i-1]$ followed by a character different from $x[i]$, $kmpNext[0] = -1$ and $kmpNext[m] = m - per(x)$.

The lists in the buckets are explicitly stored in a table $list$.

The preprocessing phase of the KmpSkip Search algorithm is in $O(m+\Sigma)$ time and space complexity.

A general situation for an attempt during the searching phase is the following (see figure 39.1):

- j is the current text position;
- $x[i] = y[j]$;
- $start = j - i$ is the possible starting position of an occurrence of x in y;
- $wall$ is the rightmost scanned text position;
- $x[0..wall - start - 1] = y[start..wall - 1]$;

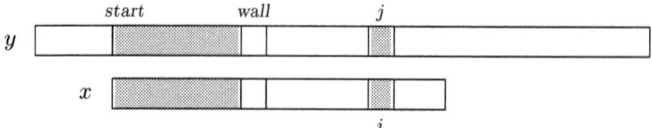

Figure 39.1. General situation during the searching phase of the KmpSkip algorithm.

The comparisons are performed from left to right between the two factors $x[wall - start\,..\,m-1]$ and $y[wall\,..\,start + m - 1]$ until a mismatch or a whole match occurs. Let $k \geq wall - start$ be the smallest integer such that $x[k] \neq y[start + k]$ or $k = m$ if an occurrence of x starts at position $start$ in y.

Then $wall$ takes the value of $start + k$.

After that the algorithm KmpSkip computes two shifts (two new starting positions): the first one according to the skip algorithm (see algorithm AdvanceSkip for details), this gives us a starting position $skipStart$, the second one according to the shift table of Knuth-Morris-Pratt, which gives us another starting position $kmpStart$.

Several cases can arise:

- $skipStart < kmpStart$ then a shift according to the skip algorithm is applied which gives a new value for $skipStart$, and we have to compare again $skipStart$ and $kmpStart$;

- $kmpStart < skipStart < wall$ then a shift according to the shift table of Morris-Pratt is applied. This gives a new value for $kmpStart$. We have to compare again $skipStart$ and $kmpStart$;

- $skipStart = kmpStart$ then another attempt can be performed with $start = skipStart$;

- $kmpStart < wall < skipStart$ then another attempt can be performed with $start = skipStart$.

The searching phase of the KmpSkip Search algorithm is in $O(n)$ time.

3 The C code

The function preMp is given chapter 6 and the function preKmp is given chapter 7.

39. KMPSKIP SEARCH ALGORITHM

```c
int attempt(String y, String x, int m, int start, int wall)
{
   int k;

   k = wall - start;
   while (k < m && x[k] == y[k + start])
      ++k;
   return(k);
}


void KMPSKIP(String x, int m, String y, int n) {
   int i, j, k, kmpStart, per, start, wall;
   int kmpNext[XSIZE], list[XSIZE], mpNext[XSIZE],
      z[ASIZE];

   /* Preprocessing */
   preMp(x, m, mpNext);
   preKmp(x, m, kmpNext);
   memset(z, -1, ASIZE*sizeof(int));
   memset(list, -1, m*sizeof(int));
   z[x[0]] = 0;
   for (i = 1; i < m; ++i) {
      list[i] = z[x[i]];
      z[x[i]] = i;
   }

   /* Searching */
   wall = 0;
   per = m - kmpNext[m];
   i = j = -1;
   do {
      j += m;
   } while (j < n && z[y[j]] < 0);
   if (j >= n) return;
   i = z[y[j]];
   start = j - i;
   while (start <= n - m) {
      if (start > wall)
         wall = start;
      k = attempt(y, x, m, start, wall);
```

```
wall = start + k;
if (k == m) {
   OUTPUT(start);
   i -= per;
}
else
   i = list[i];
if (i < 0) {
   do {
      j += m;
   } while (j < n && z[y[j]] < 0);
   if (j >= n)
      return;
   i = z[y[j]];
}
kmpStart = start + k - kmpNext[k];
k = kmpNext[k];
start = j - i;
while (start < kmpStart ||
       (kmpStart < start && start < wall)) {
   if (start < kmpStart) {
      i = list[i];
      if (i < 0) {
         do {
            j += m;
         } while (j < n && z[y[j]] < 0);
         if (j >= n) return;
         i = z[y[j]];
      }
      start = j - i;
   }
   else {
      kmpStart += (k - mpNext[k]);
      k = mpNext[k];
   }
}
      }
   }
}
```

4 The example

c	A	C	G	T
$z[c]$	6	1	7	−1

i	0	1	2	3	4	5	6	7
$list[i]$	−1	−1	−1	0	2	3	4	5

i	0	1	2	3	4	5	6	7	8
$x[i]$	G	C	A	G	A	G	A	G	
$mpNext[i]$	−1	0	0	0	1	0	1	0	1
$kmpNext[i]$	−1	0	0	−1	1	−1	1	−1	1

Searching phase

First attempt:

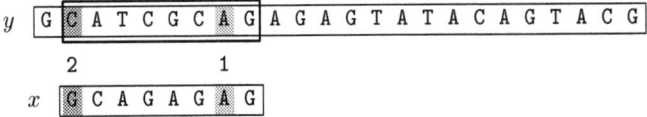

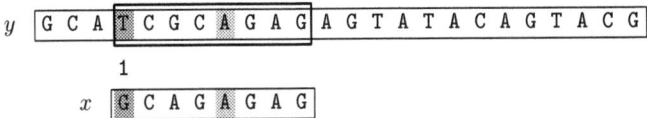

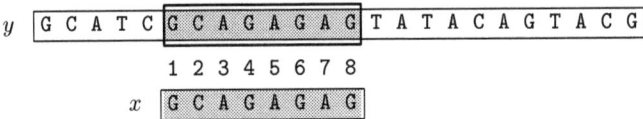

Shift by 8

Second attempt:

y G C A T C G C A G A G A G T A T A C A G T A C G

                                    1

Shift by 8

Third attempt:

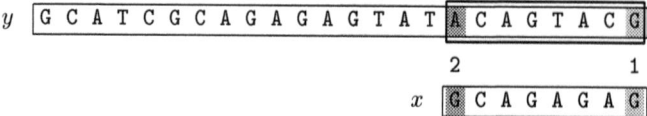

The KmpSkip Search algorithm performs 14 text character inspections on the example.

5 References

- CHARRAS, C., LECROQ, T., PEHOUSHEK, J.D., 1998, A very fast string matching algorithm for small alphabets and long patterns, in *Proceedings of the 9th Annual Symposium on Combinatorial Pattern Matching*, M. Farach-Colton ed., Piscataway, New Jersey, Lecture Notes in Computer Science 1448, pp 55–64, Springer-Verlag, Berlin.

CHAPTER 40

ALPHA SKIP SEARCH ALGORITHM

1 Main features

- improvement of the Skip Search algorithm;
- uses buckets of positions for each factor of length $\log_\sigma m$ of the pattern;
- preprocessing phase in $O(m)$ time and space complexity;
- searching phase in $O(m \times n)$ time complexity;
- $O(\log_\sigma m \times (n/(m - \log_\sigma m)))$ expected text character comparisons.

2 Description

The preprocessing phase of the Alpha Skip Search algorithm consists in building a trie $T(x)$ of all the factors of the length $\ell = \log_\sigma m$ occurring in the word x. The leaves of $T(x)$ represent all the factors of length ℓ of x. There is then one bucket for each leaf of $T(x)$ in which is stored the list of positions where the factor, associated to the leaf, occurs in x.

The worst case time of this preprocessing phase is linear if the alphabet size is considered to be a constant.

The searching phase consists in looking into the buckets of the text factors $y[j \mathinner{.\,.} j + \ell - 1]$ for all $j = k \times (m - \ell + 1) - 1$ with the integer k in the interval $[1, \lfloor(n - \ell)/m\rfloor]$.

The worst case time complexity of the searching phase is quadratic but the expected number of text character comparisons is $O(\log_\sigma m \times (n/(m - \log_\sigma m)))$.

3 The C code

The description of a linked list List can be found section 5.

```
List *z;

#define getZ(i) z[(i)]
```

```
void setZ(int node, int i) {
   List cell;

   cell = (List)malloc(sizeof(struct _cell));
   if (cell == NULL)
      error("ALPHASKIP/setZ");
   cell->element = i;
   cell->next = z[node];
   z[node] = cell;
}


/* Create the transition labeled by the
   character c from node node.
   Maintain the suffix links accordingly. */
int addNode(Graph trie, int art, int node, Character c) {
   int childNode, suffixNode, suffixChildNode;

   childNode = newVertex(trie);
   setTarget(trie, node, c, childNode);
   suffixNode = getSuffixLink(trie, node);
   if (suffixNode == art)
      setSuffixLink(trie, childNode, node);
   else {
      suffixChildNode = getTarget(trie, suffixNode, c);
      if (suffixChildNode == UNDEFINED)
         suffixChildNode = addNode(trie, art, suffixNode, c);
      setSuffixLink(trie, childNode, suffixChildNode);
   }
   return(childNode);
}


void ALPHASKIP(String x, int m, String y, int n, int a) {
   int b, i, j, k, logM, temp, shift, size, pos;
   int art, childNode, node, root, lastNode;
   List current;
   Graph trie;

   logM = 0;
   temp = m;
```

```
while (temp > a) {
   ++logM;
   temp /= a;
}
if (logM == 0)
   logM = 1;
else
   if (logM > m/2)
      logM = m/2;

/* Preprocessing */
size = 2 + (2*m - logM + 1)*logM;
trie = newTrie(size, size*ASIZE);
z = (List *)calloc(size, sizeof(List));
if (z == NULL)
   error("ALPHASKIP");

root = getInitial(trie);
art = newVertex(trie);
setSuffixLink(trie, root, art);
node = newVertex(trie);
setTarget(trie, root, x[0], node);
setSuffixLink(trie, node, root);
for (i = 1; i < logM; ++i)
   node = addNode(trie, art, node, x[i]);
pos = 0;
setZ(node, pos);
pos++;
for (i = logM; i < m - 1; ++i) {
   node = getSuffixLink(trie, node);
   childNode = getTarget(trie, node, x[i]);
   if (childNode == UNDEFINED)
      node = addNode(trie, art, node, x[i]);
   else
      node = childNode;
   setZ(node, pos);
   pos++;
}
node = getSuffixLink(trie, node);
childNode = getTarget(trie, node, x[i]);
if (childNode == UNDEFINED) {
```

```
            lastNode = newVertex(trie);
            setTarget(trie, node, x[m - 1], lastNode);
            node = lastNode;
        }
        else
            node = childNode;
        setZ(node, pos);

        /* Searching */
        shift = m - logM + 1;
        for (j = m + 1 - logM; j < n - logM; j += shift) {
            node = root;
            for (k = 0; node != UNDEFINED && k < logM; ++k)
                node = getTarget(trie, node, y[j + k]);
            if (node != UNDEFINED)
                for (current = getZ(node);
                     current != NULL;
                     current = current->next) {
                    b = j - current->element;
                    if (x[0] == y[b] &&
                        memcmp(x + 1, y + b + 1, m - 1) == 0)
                        OUTPUT(b);
                }
        }
        free(z);
}
```

4 The example

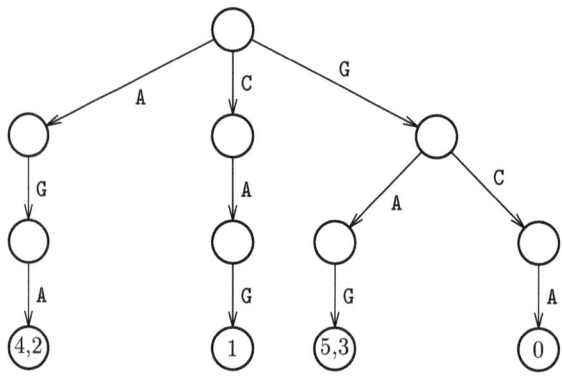

u	$z[u]$
AGA	$(4,2)$
CAG	(1)
GAG	$(5,3)$
GCA	(0)

Searching phase

First attempt:

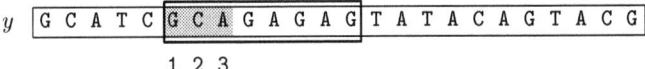

```
y  G C A T C G C A G A G A G T A T A C A G T A C G
             1 2 3
```

```
y  G C A T C G C A G A G A G T A T A C A G T A C G
             1 2 3 4 5 6 7 8
          x  G C A G A G A G
```

Shift by 6

Second attempt:

```
y  G C A T C G C A G A G A G T A T A C A G T A C G
                           1 2 3
```

Shift by 6

Third attempt:

```
y  G C A T C G C A G A G A G T A T A C A G T A C G
                                     1 2 3
```

```
y  G C A T C G C A G A G A G T A T A C A G T A C G
                                     1
                                  x  G C A G A G A G
```

The Alpha Skip Search algorithm performs 18 text character inspections on the example.

5 References

- CHARRAS, C., LECROQ, T., PEHOUSHEK, J.D., 1998, A very fast string matching algorithm for small alphabets and long patterns, in

Proceedings of the 9th Annual Symposium on Combinatorial Pattern Matching, M. Farach-Colton ed., Piscataway, New Jersey, Lecture Notes in Computer Science 1448, pp 55–64, Springer-Verlag, Berlin.

- LECROQ, T., 2000, *Quelques aspects de l'algorithmique du texte*, Habilitation thesis, Université de Rouen, France.

APPENDIX I

EXAMPLE OF GRAPH IMPLEMENTATION

A possible implementation of the interface of section 5.2 follows.

```
struct _graph {
   int vertexNumber,
       edgeNumber,
       vertexCounter,
       initial,
       *terminal,
       *target,
       *suffixLink,
       *length,
       *position,
       *shift;
};

typedef struct _graph *Graph;
typedef int boolean;

#define UNDEFINED -1

/* returns a new data structure for
   a graph with v vertices and e edges */
Graph newGraph(int v, int e) {
   Graph g;

   g = (Graph)calloc(1, sizeof(struct _graph));
   if (g == NULL) error("newGraph");
   g->vertexNumber  = v;
   g->edgeNumber    = e;
   g->initial       = 0;
   g->vertexCounter = 1;
   return(g);
}
```

```
/* returns a new data structure for
   a automaton with v vertices and e edges */
Graph newAutomaton(int v, int e) {
   Graph aut;

   aut = newGraph(v, e);
   aut->target = (int *)calloc(e, sizeof(int));
   if (aut->target == NULL)
      error("newAutomaton");
   aut->terminal = (int *)calloc(v, sizeof(int));
   if (aut->terminal == NULL)
      error("newAutomaton");
   return(aut);
}


/* returns a new data structure for
   a suffix automaton with v vertices and e edges */
Graph newSuffixAutomaton(int v, int e) {
   Graph aut;

   aut = newAutomaton(v, e);
   memset(aut->target, UNDEFINED, e*sizeof(int));
   aut->suffixLink = (int *)calloc(v, sizeof(int));
   if (aut->suffixLink == NULL)
      error("newSuffixAutomaton");
   aut->length = (int *)calloc(v, sizeof(int));
   if (aut->length == NULL)
      error("newSuffixAutomaton");
   aut->position = (int *)calloc(v, sizeof(int));
   if (aut->position == NULL)
      error("newSuffixAutomaton");
   aut->shift = (int *)calloc(e, sizeof(int));
   if (aut->shift == NULL)
      error("newSuffixAutomaton");
   return(aut);
}
```

```
/* returns a new data structure for
   a trie with v vertices and e edges */
Graph newTrie(int v, int e) {
   Graph aut;

   aut = newAutomaton(v, e);
   memset(aut->target, UNDEFINED, e*sizeof(int));
   aut->suffixLink = (int *)calloc(v, sizeof(int));
   if (aut->suffixLink == NULL)
      error("newTrie");
   aut->length = (int *)calloc(v, sizeof(int));
   if (aut->length == NULL)
      error("newTrie");
   aut->position = (int *)calloc(v, sizeof(int));
   if (aut->position == NULL)
      error("newTrie");
   aut->shift = (int *)calloc(e, sizeof(int));
   if (aut->shift == NULL)
      error("newTrie");
   return(aut);
}


/* returns a new vertex for graph g */
int newVertex(Graph g) {
   if (g != NULL && g->vertexCounter <= g->vertexNumber)
      return(g->vertexCounter++);
   error("newVertex");
}


/* returns the initial vertex of graph g */
int getInitial(Graph g) {
   if (g != NULL)
      return(g->initial);
   error("getInitial");
}
```

```
/* returns true if vertex v is terminal in graph g */
boolean isTerminal(Graph g, int v) {
   if (g != NULL && g->terminal != NULL &&
       v < g->vertexNumber)
      return(g->terminal[v]);
   error("isTerminal");
}


/* set vertex v to be terminal in graph g */
void setTerminal(Graph g, int v) {
   if (g != NULL && g->terminal != NULL &&
       v < g->vertexNumber)
      g->terminal[v] = 1;
   else
      error("isTerminal");
}


/* returns the target of edge from vertex v
   labeled by character c in graph g */
int getTarget(Graph g, int v, Character c) {
   if (g != NULL && g->target != NULL &&
       v < g->vertexNumber && v*c < g->edgeNumber)
      return(g->target[v*(g->edgeNumber/g->vertexNumber) +
                       c]);
   error("getTarget");
}


/* add the edge from vertex v to vertex t
   labeled by character c in graph g */
void setTarget(Graph g, int v, Character c, int t) {
   if (g != NULL && g->target != NULL &&
       v < g->vertexNumber &&
       v*c <= g->edgeNumber && t < g->vertexNumber)
      g->target[v*(g->edgeNumber/g->vertexNumber) + c] = t;
   else
      error("setTarget");
}
```

I. EXAMPLE OF GRAPH IMPLEMENTATION

```c
/* returns the suffix link of vertex v in graph g */
int getSuffixLink(Graph g, int v) {
   if (g != NULL && g->suffixLink != NULL &&
       v < g->vertexNumber)
      return(g->suffixLink[v]);
   error("getSuffixLink");
}


/* set the suffix link of vertex v
   to vertex s in graph g */
void setSuffixLink(Graph g, int v, int s) {
   if (g != NULL && g->suffixLink != NULL &&
       v < g->vertexNumber && s < g->vertexNumber)
      g->suffixLink[v] = s;
   else
      error("setSuffixLink");
}


/* returns the length of vertex v in graph g */
int getLength(Graph g, int v) {
   if (g != NULL && g->length != NULL &&
       v < g->vertexNumber)
      return(g->length[v]);
   error("getLength");
}


/* set the length of vertex v to integer ell in graph g */
void setLength(Graph g, int v, int ell) {
   if (g != NULL && g->length != NULL &&
       v < g->vertexNumber)
      g->length[v] = ell;
   else
      error("setLength");
}
```

```
/* returns the position of vertex v in graph g */
int getPosition(Graph g, int v) {
   if (g != NULL && g->position != NULL &&
       v < g->vertexNumber)
      return(g->position[v]);
   error("getPosition");
}


/* set the length of vertex v to integer ell in graph g */
void setPosition(Graph g, int v, int p) {
   if (g != NULL && g->position != NULL &&
       v < g->vertexNumber)
      g->position[v] = p;
   else
      error("setPosition");
}


/* returns the shift of the edge from vertex v
   labeled by character c in graph g */
int getShift(Graph g, int v, Character c) {
   if (g != NULL && g->shift != NULL &&
       v < g->vertexNumber && v*c < g->edgeNumber)
      return(g->shift[v*(g->edgeNumber/g->vertexNumber) +
             c]);
   error("getShift");
}


/* set the shift of the edge from vertex v
   labeled by character c to integer s in graph g */
void setShift(Graph g, int v, Character c, int s) {
   if (g != NULL && g->shift != NULL &&
       v < g->vertexNumber && v*c <= g->edgeNumber)
      g->shift[v*(g->edgeNumber/g->vertexNumber) + c] = s;
   else
      error("setShift");
}
```

I. EXAMPLE OF GRAPH IMPLEMENTATION

```
/* copies all the characteristics of vertex source
   to vertex target in graph g */
void copyVertex(Graph g, int target, int source) {
   if (g != NULL && target < g->vertexNumber &&
       source < g->vertexNumber) {
      if (g->target != NULL)
         memcpy(g->target +
                target*(g->edgeNumber/g->vertexNumber),
                g->target +
                source*(g->edgeNumber/g->vertexNumber),
                (g->edgeNumber/g->vertexNumber)*
                sizeof(int));
      if (g->shift != NULL)
         memcpy(g->shift +
                target*(g->edgeNumber/g->vertexNumber),
                g->shift +
                source*(g->edgeNumber/g->vertexNumber),
                g->edgeNumber/g->vertexNumber)*
                sizeof(int));
      if (g->terminal != NULL)
         g->terminal[target] = g->terminal[source];
      if (g->suffixLink != NULL)
         g->suffixLink[target] = g->suffixLink[source];
      if (g->length != NULL)
         g->length[target] = g->length[source];
      if (g->position != NULL)
         g->position[target] = g->position[source];
   }
   else
      error("copyVertex");
}
```

INDEX

ASIZE, 6
XSIZE, 6
YSIZE, 6
Character, 5
OUTPUT, 6
String, 6

Aho, A.V., 26, 35, 41, 88, 121
Allauzen, C., 172
alphabet, 1
Aoe, J.-I., 41, 88
Apostolico, A., 71, 110
attempt, 2

Baase, S., 41, 89
bad-character shift, 83, 100, 119, 127, 135, 139, 143, 147, 151, 201, 207
Baeza-Yates, R.A., 20, 26, 30, 41, 42, 89, 90, 121, 160
basic, 5
Beauquier, D., 35, 41, 51, 89, 122
Berry, T., 146
Berstel, J., 35, 41, 51, 89, 122
Binstock, A., 89
bitwise techniques, 27
border, 5, 31
Boyer, R.S., 89
Breslauer, D., 60, 65, 193

Cantone, D., 126
Cardon, A., 77
Charras, C., 77, 214, 220, 225
Chrétienne, P., 35, 41, 51, 89, 122
Cole, R., 89
Colussi, L., 60, 118

Cormen, T.H., 20, 26, 41, 89
critical factorization, 187
Crochemore, M., 20, 26, 30, 35, 41, 42, 51, 71, 82, 89, 103, 110, 122, 129, 160, 165, 166, 172, 185, 193, 200
Czumaj, A., 103, 160, 166

delay, 32, 38, 46

factor, 4
factorization, 187
Faro, S., 126
Flajolet, P., 42

Gąsieniec, L., 103, 160, 166
Galil, Z., 60, 65, 93, 185
Giancarlo, R., 60, 65, 110
Gonnet, G.H., 20, 26, 30, 42, 90
good-suffix shift, 83, 99, 201, 207
Goodrich, M.T., 42, 90
Gusfield, D., 42, 90, 111

Hancart, C., 20, 26, 30, 35, 41, 42, 51, 52, 71, 77, 89, 90, 103, 110, 122
hashing function, 21
holes, 53
Hopcroft, J.E., 35
Horspool, R.N., 122
Hume, A., 138

Jarominek, S., 103, 160, 166

Karp, R.M., 26
Knuth, D.E., 42, 90

Lecroq, T., 20, 26, 30, 35, 41, 42, 51, 89, 90, 93, 97, 103, 110, 111, 122, 129, 160, 166, 214, 220, 225, 226
Leiserson, C.E., 20, 26, 41, 89
local period, 187

Manber, U., 30
matching shift, 83
maximal suffix, 188, 195
Maximal-Suffix decomposition, 195
Moore, J.S., 89
Morris, Jr, J.H., 35, 42, 90

Navarro G., 30, 41, 89, 160
Navarro, G., 30, 42, 90, 122, 160, 172, 177
nohole, 61
noholes, 53
non-periodic, 5

oracle, 167

pattern, 1
Pehoushek, J.D., 214, 220, 225
perfect factorization, 179
period, 5
periodic, 5, 162
Perrin, D., 193
Plandowski, W., 103, 160, 166
Pratt, V.R., 35, 42, 90
prefix, 4
prefix period, 179

Rabin, M.O., 26
Raffinot M., 172
Raffinot, M., 30, 42, 90, 122, 160, 172, 177
Raita, T., 154
Ravindran, S., 146
repetition, 187
reverse, 5

Rex, J., 89
Ribeiro-Neto B., 30, 41, 89, 160
Rivest, R.L., 20, 26, 41, 89
Rytter, W., 26, 35, 42, 51, 82, 89, 103, 110, 160, 166, 185, 193, 200
Régnier, M., 121

Sedgewick, R., 26, 42, 90
Seiferas, J., 185
shift, 2
Simon, I., 52
sliding window mechanism, 2
Smith, P.D., 149, 154
Smyth, W. F., 26, 30, 43, 71, 90, 93, 97, 103, 122, 130, 205, 210
Stephen, G.A., 26, 43, 90, 93, 122, 130, 138
String matching, 1
substring, 4
subword, 4
suffix, 4
suffix automaton, 79, 155, 162
suffix oracle, 167
Sunday, D.M., 130, 138, 205, 210

tagged border, 37
Takaoka, T., 142
Tamassia, R., 42, 90
Tamm, M., 134
text, 1
the local period, 187
the period, 5
trie, 221
turbo-shift, 99

Ullman, J.D., 35

Van Gelder, A., 41, 89

Watson, B.W., 43, 90

window, 1
Wirth, N., 43, 90
Wu, S., 30

Yao, A.C., 160, 166

Zhu, R.F., 142

Christian Charras
Département d'Informatique
Faculté des Sciences et des Techniques
Université de Rouen
76821 MONT-SAINT-AIGNAN CEDEX – FRANCE
Email: Christian.Charras@univ-rouen.fr

Thierry Lecroq
ABISS
Faculté des Sciences et des Techniques
Université de Rouen
76821 MONT-SAINT-AIGNAN CEDEX – FRANCE
Email: Thierry.Lecroq@univ-rouen.fr

www.ingramcontent.com/pod-product-compliance
Ingram Content Group UK Ltd.
Pitfield, Milton Keynes, MK11 3LW, UK
UKHW021318180426
11947UKWH00015B/1304